T0211844

Neurodiversity Studies

Building on work in feminist studies, queer studies, and critical race theory, this volume challenges the universality of propositions about human nature, by questioning the boundaries between predominant neurotypes and 'others', including dyslexics, autistics, and ADHDers.

This is the first work of its kind to bring cutting-edge research across disciplines to the concept of neurodiversity. It offers in-depth explorations of the themes of cure/prevention/eugenics; neurodivergent wellbeing; cross-neurotype communication; neurodiversity at work; and challenging brain-bound cognition. It analyses the role of neuro-normativity in theorising agency, and a proposal for a new alliance between the Hearing Voices Movement and neurodiversity. In doing so, we contribute to a cultural imperative to redefine what it means to be human. To this end, we propose a new field of enquiry that finds ways to support the inclusion of neurodivergent perspectives in knowledge production, and which questions the theoretical and mythological assumptions that produce the idea of the neurotypical.

Working at the crossroads between sociology, critical psychology, medical humanities, critical disability studies, and critical autism studies, and sharing theoretical ground with critical race studies and critical queer studies, the proposed new field – neurodiversity studies – will be of interest to people working in all these areas.

Hanna Bertilsdotter Rosqvist is an Associate Professor in Sociology and currently a Senior Lecturer in Social work at Södertörn University. Her recent research is around autism, identity politics, and sexual, gendered, and age normativity. She is the former Chief Editor of *Scandinavian Journal of Disability Research*.

Nick Chown is a book indexer who undertakes autism research in his spare time. His recent publication is *Understanding and Evaluating Autism Theory* (2016). He has reviewed for the *Journal of Autism and Developmental Disorders*, written various academic articles, and led a team studying support for autistic students in the UK.

Anna Stenning is a Wellcome Trust Research Fellow in the Humanities and Social Sciences at the University of Leeds. Her current research focuses on literary representations of autism, and intersections between disability studies and environmental justice. She is also a co-editor, with David Borthwick and Pippa Marland, of *Walking, Landscape and Environment* (2019).

Routledge Advances in Sociology

For more information about this series, please visit: www.routledge.com/
Routledge-Advances-in-Sociology/book-series/SE0511

Neurodiversity Studies

A New Critical Paradigm

Edited by Hanna Bertilsdotter Rosqvist, Nick Chown and Anna Stenning

Routledge
Taylor & Francis Group

LONDON AND NEW YORK

First published 2020
by Routledge
2 Park Square, Milton Park, Abingdon, Oxon OX14 4RN

and by Routledge
52 Vanderbilt Avenue, New York, NY 10017

Routledge is an imprint of the Taylor & Francis Group, an informa business

British Library Cataloguing-in-Publication Data
A catalogue record for this book is available from the British
Library

Library of Congress Cataloging-in-Publication Data
A catalog record has been requested for this book

ISBN: 978-0-367-33831-2 (hbk)
ISBN: 978-0-367-50325-3 (pbk)
ISBN: 978-0-429-32229-7 (ebk)

Typeset in Times New Roman
by Wearset Ltd, Boldon, Tyne and Wear

'My [autistic] personhood is intact. My selfhood is undamaged. I find great value and meaning in my life, and I have no wish to be cured of being myself' (Sinclair 1992, p. 302).[1]

Note

1 Sinclair, J. (1992). Bridging the gaps: An inside-out view of autism. In: E. Shopler & G. B. Mesibov (Eds.), *High-functioning individuals with autism* (pp. 294–302). Boston, MA: Springer.

Contents

Contributors

Matthew K. Belmonte is a neuroscientist, and the brother and uncle of two people with autism. His research explores the brain physiology that underlies autistic cognitive traits in people with autism spectrum conditions, in their family members, and in the general population. His work has shown that some of the same factors that produce autism also underlie human cognitive diversity in general; in this sense, people with autism can be said to be 'human, but more so'. Belmonte has worked within (and in a couple of cases been run out of!) institutions including the University of California San Diego, the University of Cambridge, Cornell University, and the National Brain Research Centre (India). He is the recipient of a 2009 US National Science Foundation Faculty Early Career Development Award, the 2010 Neil O'Connor Award from the British Psychological Society, a 2011 Fulbright-Nehru fellowship, and a 2018 NHS Research for Patient Benefit grant.

Virginia Bovell was a social policy researcher at London School of Economics before giving up paid work to care for her autistic son. She has campaigned and written on a range of education, social, and policy issues with regard to autism and learning disability. In 2015 she was awarded a doctorate by Oxford University for her research into the ethical challenges posed by attempts to prevent and/or cure autism.

Robert Chapman is currently a Vice-Chancellor's Research Fellow at the University of Bristol, where he leads the Health and Wellbeing for a Neurodiverse Age project. Prior to this he taught at the University of Bristol and Kings College London, and wrote his doctoral thesis on the ethics of autism at the University of Essex.

Dennis Hansson is a journalist.

Akiko Hart is the CEO of the National Survivor User Network. She has previously worked as the Hearing Voices Project Manager at Mind in Camden and the Director of Mental Health Europe. She is the Chair of ISPS UK and a Committee Member of the English Hearing Voices Network and National Voices. She sits on the Advisory Board of Durham University's Institute for Medical Humanities.

Serena Hasselblad has a doctorate in Technology from Chalmer University, Sweden. She works with autistic peer support and lectures about autism.

Alyssa Hillary is an autistic PhD student in the Interdisciplinary Neuroscience Program at the University of Rhode Island. Officially, they work on augmentative and alternative communication, both in the form of brain computer interfaces and as used by autistic adults. Unofficially but more prolifically, they write about issues related to neurodiversity and representation in fiction, media, scientific research, and life writing. Their work has appeared in *Criptiques, Disability in Kidlit, International Perspectives on Teaching with Disability*, and in several Autonomous Press anthologies.

Inês Hipólito works on the intersection between philosophy of mind and Neuroscience. Her specialisation is in the architecture and organisation of cognition and mental phenomena. Hipólito's work investigates the relations between the classic modularity of the mind and more recent computational theories under the Bayesian brain. This requires combining research from philosophy of mind and epistemology with inference theory under network and dynamic modelling. Hipólito has co-edited a special issue for *Philosophical Transactions* (2019). She has also co-edited a couple of volumes for the Mind and Brain Studies (Springer) exploring the extended and embodied mind on the topics of mind-technology interaction (2020), and philosophy of psychiatry (2018). She has published work in edited books (Routledge, CUP) and journals (*Australasian Philosophical Review, Physics of Life Reviews, Journal of Clinical Practice, Progress in Biophysics and Molecular Biology*). Hipólito's work has been honoured with prizes and awards, including the Berlin School of Mind and Brain; the University of Oxford; the Federation of European Neuroscience Societies; and the British Association for Cognitive Neuroscience. She has been invited to give talks and lectures at the University of Cambridge, the University of Oxford, University of Sussex.

Dieuwertje Dyi Huijg recently completed her PhD in Sociology, at the University of Manchester. Her thesis concerns an empirically grounded theoretical exploration of 'intersectional agency', which she identifies as a polylithic mechanism. Dyi is currently working on a new project on intersectionality, agency, resistance, and ableism and neuronormativity.

Daniel D. Hutto is Senior Professor of Philosophical Psychology and Head of the School of Liberal Arts at the University of Wollongong. He is co-author of the award-winning *Radicalizing Enactivism* (MIT, 2013) and its sequel, *Evolving Enactivism* (MIT, 2017). His other recent books, include: *Folk Psychological Narratives* (MIT, 2008) and *Wittgenstein and the End of Philosophy* (Palgrave, 2006). He is editor of *Narrative and Understanding Persons* (CUP, 2007) and *Narrative and Folk Psychology* (Imprint Academic, 2009). A special yearbook, *Radical Enactivism*, focusing on his philosophy of intentionality, phenomenology, and narrative, was published in 2006. He is regularly invited to speak not only at philosophy conferences but at expert meetings of anthropologists, clinicians, educationalists, narratologists, neuroscientists, and psychologists.

David Jackson-Perry is a doctoral candidate at Queen's University, Belfast, UK, and a specialist in sexual health. His research centres around an exploration of autistic experiences of sexuality and intimacy, with a particular focus on sensory needs and desires.

Alan Jurgens is a PhD candidate at the University of Wollongong and has an MA in philosophy with a specialisation in phenomenology from the University of Copenhagen. His current research focuses on the field of social cognition, specifically the debate between the traditional cognitivist approach (theory of mind) and alternative enactivist explanations. In addition to this, he has researched and presented on Autism Spectrum Disorder, false-belief tasks, and social robots at the Wollongong Social Robotics Workshop (2017). In this area, he is primarily interested in addressing the question of whether making false-belief tasks more interactive and reducing social cues through the use of robots can lead to improved performance by those diagnosed with Autism Spectrum Disorder.

Marianthi Kourti is an autistic doctoral candidate at the University of Birmingham, UK, a specialist mentor for autistic students and a former SEN teacher. Their research is looking into how autistic individuals form a sense of gender identity, with a particular interest in using the autistic experience to interrogate norms of any kind.

Jenn Layton Annable is a fellow of the Institute of Mental Health, University of Nottingham, UK, and an independent autistic autism researcher investigating the intersection between gender, autistic experience, and self-identity. She has profound differences inner sensory experiences that have affected her life and continue to shape her work and future perspective.

Nicola Martin is Head of Research and Higher Degrees at London South Bank University (LSBU) in The Division of Education. She has held various senior academic and professional roles in education. These include heading the disability and wellbeing service at the LSE and the Autism Centre at Sheffield Hallam. Nicola gained a National Teaching Fellowship for work on inclusive education and universal design for learning. She has a track record of social model focused participatory autism and disability research, including The Cambridge Asperger's Project with Professor Simon Baron-Cohen. Nicola heads the Critical Autism and Disability Studies (CADS) Research Group at LSBU. The Participatory Autism Research Collective (PARC) was born within CADS. Nicola is very clear that autism research from CADS will always involve properly remunerated autistic researchers. PARC has turned into a national influencer which aims to progress this agenda. The timeline of Nicola's career is largely behind her and her main concern at this stage is to encourage and support other people in their development.

Kirke Nilsson is a master's student and lecturer in Social Work at Örebro university, Sweden. Her research centres around work life and autism.

Linda Örulv is a researcher at Linköping University, Sweden. Her research centres around people with dementia.

Hajo Seng is a doctoral candidate in Education at the University of Hamburg. His research focuses on exploration of autistic experiences and ways of thinking. He works with autistic peer support.

Mitzi Waltz is a researcher in Global Health at the Athena Institute of Vrije Universiteit Amsterdam, and a senior researcher with Disability Studies in Nederland. Previously, she was Senior Lecturer in Autism Studies at Sheffield Hallam University and Lecturer in Autism Studies at the University of Birmingham in the UK. She is the author of *Autism: A Social and Medical History* (Palgrave Macmillan, 2013).

Acknowledgements

The editors are indebted to the autism community, autism self-advocates, and neurodiversity activists, without whom the idea for this book would not have been possible. It has taken many years of ardent campaigning by those associated with the Autism Self-Advocacy Network, PARC, and DANDA for autism, ADHD, dyslexia, and other long-term neurological differences to be even considered to be part of a broader spectrum of human diversity, rather than inescapably associated with deviance, disorder, or impoverished selfhood. We believe that a position of neurodiversity is essential to combat the stigma that undermines the wellbeing of neurodivergent individuals, while at the same time we need to pay greater attention to combinations of neurological differences that can prove inherently difficult to manage within our increasingly homogenised and individualised world.

In our attempt to advance the theorisations of these activist positions we are very grateful to our authors who draw on their expertise in cognitive science, the philosophy of psychiatry, sociology, critical autism studies, neuroscience, the humanities, ethics, and disability studies with critical insight and personal dedication. Not all authors who have contributed to this work identify as neurodivergent, but we count on their support as allies. We also thank our families and friends for supporting our work and putting up with a work that has become a labour of love. Finally, we would like to thank our editors at Routledge for believing in this project.

Introduction

Hanna Bertilsdotter Rosqvist, Anna
Stenning and Nick Chown

Scholars have now 'amassed overwhelming evidence of the extent to which the myths of the ideal rational person and the 'universality' of propositions about human nature have been oppressive to those who are not European, white, male, middle class, Christian, able-bodied, thin, and heterosexual' (Ellsworth, 1989, p. 304). The concept of neurodiversity usually refers to perceived variations seen in cognitive, affectual, and sensory functioning differing from the majority of the general population or 'predominant neurotype', more usually known as the 'neurotypical' population. The ontological assumption of neurodiversity is often contrasted with the ideological position that there is a recurrent 'normal' cognitive, affective, and sensory type, otherwise known as the normal human being or the 'normate' (Garland Thomson, 1997), as defined in the first half of the twentieth century, by way of the prerogative of psychologists who were living in the shadow of eugenics. The attribute of neurodivergence may also be applied – more problematically – to individual people. As Nick Walker says, 'Diversity is a trait possessed by a group, not an individual' (2014) and to talk of individuals as neurodiverse is to situate them as 'other' to the norm (as well as being, in our opinion, nonsensical).

The construct of the neurotypical as an ideal ethical subject may be added to the list of assumptions about ideal rationality, since rationality is often conceived in terms of cognitive, social, and sensory 'behaviours' – particularly within the cognitive-behavioural paradigm in psychology. Through describing behaviours from the outside, and through medical, economic, and social interventions, subjects are being produced as deviants to assumed standards of intellectual, perceptual, and emotional processing. They become subjects to both internal and external oppression. This ideal of rationality measured through external behaviours is often exclusionary of those who may be able to offer alternative conceptualisations. This book seeks to demonstrate the effects of this exclusion, and to begin to address it through our position as neurodivergent – or allied – 'insiders' in the academic realm. While not all our contributors identify as neurodivergent, we share a commitment to 'decentring' the cognitive, affectual, and sensory norm.

Those who share a form of neurodivergence – such as bipolar or hearing voices – may be referred to as a 'neurominority'. We use the expressions 'neurodivergence' and 'neurodivergent conditions' simultaneously and as alternatives to the idea of 'persons with disorders' (conceived as an inherently harmful condition that impacts on a person's flourishing) that is encompassed by standard medical models of conditions such as autism and dyslexia.[1] This is similar to the way in which autism self-advocates have, since the 1990s, argued for identity-first language in the discussion of autism. In this way, we are concerned with providing a new theorisation of conditions that are understood as impacting on the individual's sense of identity, alongside differences from standard forms of perceiving and responding to the world. This contrasts with the search to find neurological differences that are perceived as an 'interruption' (permanent or otherwise) from an individual's proper life. Whether the medical conditions coincide with these 'subjectively experienced' differences and identities is a matter of empirical investigation. For this reason, the scoping of neurodiversity remains, as we explore in the conclusion, something that needs further reflection.

This book aims to problematise neurotypical domination of the institutions and practices of academic knowledge production, by questioning the boundaries between the predominant neurotypes and their 'others'. We address the cultural and social processes of 'cognitive othering' through which they have arisen in the West during the past 100 years. The creation of boundaries – both by the neurotypical and the neurodivergent – is always subject to cultural and ideological pressures. By using neurodivergence as a position with epistemological and ethical implications we will decentre – if temporarily – dominant perspectives.

With this in mind, the contributors in the book seek to formulate alternative perspectives on cognitive normativity and cognitive othering from a neurodivergent perspective including, but not limited to, autistic perspectives. We are working at the crossroads between sociology, critical psychology, critical medical humanities, disability studies, and critical autism studies to expand each of these fields, but also to define a new field of enquiry: neurodiversity studies. This field addresses the epistemic and ideological rules that govern and produce 'normals' and 'others' according to scientific, cultural, and social practices. We see this as an open and iterative process as new forms of cognitive divergence are recognised, and scientific and cultural developments provide further evidence of the diverse biopsychosocial configurations of the human. For too long, the empirical sciences – by way of the dominance of cognitive and developmental psychology, with the anticipated validation of neuroscience – has had the final say on what it is to be human. By de-pathologising neurodivergence and the identities of neurodivergent people, it is possible that what are currently known as autism, ADHD, Tourettes, dyslexia, hearing voices, bipolar disorder, Down syndrome, and dementia, may be recognised as components on a broader continuum of sensory, affectual, and cognitive processing. And as we'll see later, this still requires the recognition that particular configurations of human minds come with particular challenges within different stages of our lives, including those that are considered impairments by the individual. We believe that this position on cognitive or neurological divergence coincides with the enactive approach to the mind and allows for a greater recognition of how we can support our collective human flourishing (for further discussion, see Jurgens' chapter in this book).

While we are relegated to the margins of academic knowledge production or included as a mere gesture to be included under diversity incentives, we will struggle to have our voices heard. This risks 'othering' those who may be oppressed according to other forms of difference such as race, class, or sexuality – but who are or consider themselves to be neurotypical. However, informed by insights developed within neuro-queer theory and crip theory, we propose here to create a space to question norms governing sensory, affectual, and cognitive functioning.

For this purpose, the book aims to begin to challenge and turn neurotypical-dominated scholarship on autism and other forms of neurodivergence on its head in the same way that feminist scholarship has challenged the dominance of male perspectives. But to do so, new theories and new research methods need to be developed, refined, and transmitted. Some of them are starting to unfold in the following chapters. As intersectional feminist discourses today are critiquing and moving beyond earlier feminist ideas to include people of colour, class issues, LGBT identities, etc., fields such as critical race studies and queer theory are recognising neurodivergence as a necessary intersection.

We are not claiming to be the first to practice this move, and in the chapters that follow it becomes clear that we are following in the footsteps of the pioneers of autism rights, autistic self-advocacy, and neurodiversity theorising and activism. Our efforts to use our critical skills to question the assumption of neurotypicality and dominant constructions of rationality and sensory processing – alongside our efforts to reposition neurodivergent experiences and processes – are just a small part of the possibilities of 'neuroqueering' (see Walker, 2015; Yergeau, 2018). The adjective 'neuroqueer' – like 'neurodivergent' – is a social identity that, perhaps more helpfully, is developed through intentional practices rather than a medical diagnosis. In *Authoring Autism: On Rhetoric and Neurological Queerness*, Yergeau points out the material and discursive overlaps between queerness and cognitive difference in social and clinical practices in the US (2018). According to her own employment of 'crip-queer logic', Yergeau sees neuroqueering as a way of being in the world that 'uncovers, upsets, and unsettles power structures in normative spaces' (p. 205).

Through a focus on the possibilities of queering the dominant attribution of demi-rhetoricity to autistic people, Yergeau shares methodological ground with Alison Kafer whose *Feminist, Queer, Crip* explores both the temporality of distinctions between 'able-bodied' and 'non-disabled' people and the ways that ideas of futurity are invoked in unhelpful ways to disabled people. She explains:

> I, *we*, need to imagine crip futures because disabled people are continually being written out of the future, rendered as the sign of the future nobody wants [...] we must imagine futures that include all of us.
>
> (2013, p. 46)

Kafer's preference for the term crip – and relatedly, the fields of crip theory and crip studies – over disability and disability studies, originates in what she calls, following McRuer (2006), the 'contestatory' nature of its basis in identity politics (p. 15). While the reference to cripping and queerness may alienate those outside of the cultures in which they have been reappropriated, they are used by Kafer and Yergeau in non-exclusionary ways, and offer a deeply politicised language with which we may question compulsory able-bodiedness, and the related concept of (compulsory) able-mindedness (Ward & Price, 2016) as well as cognitive, affective, and sensory 'normates' (for further discussion, see the chapter by Jackson-Perry, Bertilsdotter Rosqvist, Kourti, and Annable in this book). Kafer (2013, p. 32) draws on the work of José Esteban Muñoz to acknowledge the ways in which disability intersects within and alongside matters of race, gender and sexuality, and class.

Setting the scene: meanings of neurodiversity and related concepts

Neurodiversity is commonly described as a reaction and challenge to existing 'medical' models of disability (also referred to as a deficit model or individual model) that describes autism and other forms of neurodivergence (as with any physical 'handicaps') as neurological defects/disorders that reside exclusively in the individual. By the medical model, a compassionate and just society should invest resources to attempt to cure or ameliorate all neurological differences medically or to improve 'functioning', where functioning is defined by both statistical measures and cultural ideals such as independence, economic productivity, and sociability.

The aim of the medical model is thus twofold: to prevent an individual from living a life that deviates from a supposedly ideal state and to ameliorate those difficulties that arise from living in a society that is constructed according to assumptions of the ideal neurological state. The second aim is compatible with the ambitions of some proponents of the neurodiversity paradigm, particularly where devices, legal protections, or medication can be helpful in tackling some of the impairments that result from societal demands, as long as these are consistent with some possibility of self-determination. Under the medical model, 'situated' or lived experience by neurodivergent individuals – so-called 'patient accounts' or 'first-hand accounts' – are only relevant in so far as they assist in the attainment of medical or age-appropriate social outcomes (such as educational or employment outcomes), and determine whether the patient or research participant has been 'cured' or can pass as such, sufficiently to become or be economically productive. Given that the medical model sees 'selfhood' as problematic in the case of neurological difference, it is at least in some ways antithetical to neurodivergence as a valid form of human identity.

In direct opposition to the medical model of disability, and relevant to those produced by what are considered neurological disorders, the social model of disability is concerned with 'all the things that impose restrictions on disabled people; ranging from individual prejudice to institutional discrimination, from inaccessible public buildings to unusable transport systems, from segregated education to excluding work arrangements, and so on' (Oliver, 1996, p. 33). The social model demands that society removes those barriers. Barriers – both attitudinal and physical – restrict life choices for those who would otherwise be disabled by the social and cultural responses to impairments – defined here as literally a 'loss' in some ability that is assumed to be common to humans – as well as the impairments themselves.[2] According to the social model of disability, society should ensure that all individuals with neurological or physical differences can be independent and equal in society, with choice and control over their own lives. This may also involve medical intervention to assist with what may be perceived as harmful to the individual. Under the social model, accounts of lived experience are essential as they are the best guide for researchers on barriers to independence and equality. As with the medical model of neurological difference, there is little recognition of the ways that embodied differences may bring about new forms of knowledge and potential in individual or family lives (for further discussion, see Jurgens' chapter in this book).

As there are no exhaustive definitions of neurodiversity or neurodivergence, there are no definitive scopings or definitions of the concept of neurotypicality, which is conceptually dependent on these. From the 'outside' – as in when it is inflected by medical-model views – it is defined as follows:

> [A] [d]escription of a medically and psychologically healthy individual demonstrating a normative pattern of neurodevelopment. Typically used specifically in contrast with individuals experiencing an atypical developmental course, such as autism.
>
> (Perszyk, 2013, p. 1)

From the perspective of neurodiversity activism, a non-normative pattern of neurodevelopment is not 'psychologically unhealthy'. Neurodivergent conditions simply represent the wide variety of differences among humanity, and which entail particular epistemological and ethical connotations. But the reference to atypical development in Perszyk's definition captures one view of autism and other conditions often included in lists of those who are said to be neurodivergent. So-called cognitive 'normality' arises as part of a discourse of normative patterns of development. This reflects a 'developmental' view of neurodivergence originating in developmental psychology, which has itself come in for criticism (for further discussion, see Chapter 1). Other conditions – such as schizophrenia – which are considered by some scholars to be aspects of neurodiversity are seldom considered developmental in nature. The concept of neurodiversity is also broadly – and not unproblematically – aligned with environmental discourses. It has been described in terms of an analogy with E. O. Wilson's concept of 'biodiversity', where various types of human minds are considered as important for the survival of the human species as biological diversity of distinct species has been seen as essential to life in general (Blume, 1997; Singer, 1999).

From another perspective, neurodiversity is most significantly a social category, as part of an argument for social justice, whereby a politics of neurological diversity recognises power inequalities between people differently situated in relation to neurology, comparable with social stratifications such as class, gender, and ethnicity (Singer, 1999). From this position, we will be interested in the lived experiences of neurodivergent people for what they tell us about problems with social and cultural worlds, and how these can and should be different.

Meanings of neurodiversity also gain force from political movements in different national and global contexts, which are commonly associated with the autistic self-advocacy or 'autism rights' movements (Baron-Cohen, 2015). We have chosen to briefly explore, here, some of the background to political inflections of neurodiversity that have emerged in recent decades, broadly outlining two different orientations within the neurodiversity movements.

The first orientation is marked by external debates aimed at reconceptualising meanings of autism, often in opposition to the groups of parents of autistic children and professionals seeking a cure for autism. This first orientation has challenged parent-dominated movements and research and politics not informed by autistic people, stressing that autism involves neurological difference (Broderick & Ne'Eman, 2008) rather than disorder. Parent-dominated autism movements (Orsini, 2009) have been criticised for standing in the way of more radical formulations of autistic politics (Chamak, 2010), and their right to represent autism and autistic people has been challenged (Bumiller, 2008, p. 968). The deficit model of autism is questioned (Chamak, 2010), and neuro-equality is promoted. The promotion of neuro-equality includes challenging 'extant social institutions that either expressly or inadvertently model a social hierarchy where the interests or needs of individuals are ranked relative to what is regarded as properly functioning cognitive capacities' (Fenton, 2007). It stresses that there is 'no single way to be "normal"' (Baron-Cohen, 2015); and may coincide with an 'ability' or strength model of autism (Chamak, 2010). Autism is also celebrated as an inseparable aspect of identity (Kapp, Gillespie-Lynch, Sherman, & Hutman, 2013; Bumiller, 2008), demanding 'recognition and acceptance' (Jaarsma & Welin, 2012). These positions are vulnerable to accusations that could be made against any form of identity politics and particularly those that involve disability or impairments but have often paved the way for genuine improvements in autistic people's senses of self-worth. Neurodiversity movements have also been noted for their alternative approach to political organisation and social action involving mainly online organising, radical repudiation of conventional social norms, and unusual political sensibilities (cf. Bumiller, 2008). These positions may contribute to the 'cognitive othering' of different forms of neurodiversity, where no obvious 'benefit' can be found.

The second orientation is associated with internal debates within neurodiversity movements and is centred around meanings of autistic or neurodivergent experience and the questioning of normalcy both within the movements and outside. On one hand, it questions excluding practices within the movement(s) (Hughes, 2015). It points at risks of reproduction of an 'us–them' attitude to neurotypicals, and for reproducing neoliberal agendas and subjects, while not offering a challenge to commodification of difference (Runswick-Cole, 2014). On the other hand, rather than looking for recognition of neurological divergence, the value of working towards 'a society that is grounded in neurological pluralism' (Perry, 2011) and 'dislodging the hegemonic dominance of what constitutes "normalcy"' (Milton, 2014b) has been advocated.

Based on these ideas formulated within neurodiversity movements, scholars have argued for the importance of including autistic experiential knowledge both in research design and implementation, and in support. In this way, they stress the importance of understanding autism and providing insights for professionals working with autistic people (Perry, 2012). This aims at both more respectful treatment (Nicolaidis, 2012), policy reforms to build stronger support systems (Ong, 2014), and new roles for professionals in promoting self-advocacy among autistic people (Ong, 2014). Ideas formulated within the neurodiversity movements have also impacted the reformulation of other social categories such as gender and sexual subjectivities (Bumiller, 2008), meanings of citizenship (Bumiller, 2008), and of the redistributive aims of the welfare state (Orsini, 2012).

The first orientation described earlier, aiming at challenging deficit models of autism by way of assuming an evolutionarily valuable continuum of neurological difference, stressed the importance of 'first-hand narratives'. But it did not question the epistemological grounds of cognitive normate privileges and epistemological authority (and legitimacy) in knowledge production. From the second orientation, the importance of both new research methods and theories has been central. For example, Milton (2014a) has called for the involvement of autistic scholars in autism research if such research is to achieve epistemological and ethical integrity.

Both neurodiversity orientations tacitly assume neurodivergence as a potentially valuable form of human existence. Acknowledging neurological difference does not imply that all difference is good in itself, or that human traits associated with neurodivergence are always desirable, but it accepts that there are 'good' and 'not so good' traits in all human beings. Both neurodiversity orientations have an affirmative perspective on neurodivergence in common, in contrast to deficit-based models that support efforts to prevent and cure individuals or find ways to ameliorate their differences and include them in 'normal' society. However, we want to stress that an anti-cure/prevention approach does not imply a rejection of the use of interventions for those neurodivergent individuals for whom interventions will be beneficial. Generally speaking, neurodiversity perspectives do not exclude the use of all interventions, only those that seek to 'normalise' an individual in a damaging and futile attempt to produce a person closer to a neurotypical 'ideal type' or by removing traits that are disliked by neurotypicals but benefit the neurodivergent individual. Nor does the anti-cure/prevention perspective necessarily preclude acceptance of parental choice pre-birth based on biomarkers. New scientific discoveries about biomarkers for autism and other neurodivergent conditions such as ADHD and dyslexia, alongside the lived experiences and reflections of neurodivergent scholars and activists, will no doubt pave the way for new theoretical perspectives in neurodiversity studies. We believe it is urgent that scholars do so, before the urge to render humanity free of deficits becomes an imperative to reduce society of the burden of disability (in this case, in the form of neurodiversity), before its full implications are understood.

Introducing the chapter content

We have chosen to organise the chapters in five different themes. In the first theme, **Curing neurodivergence/eugenics**, Mitzi Waltz, Nick Chown, and Virginia Bovell contribute to explorations into the production of the category of autism and neurodivergent subjects through processes of cognitive othering. In her chapter, Mitzi Waltz focuses on the historical development of psychological measures for normal subjects in the aftermath of World War I. Presenting a neurodivergent perspective to the field of critical psychiatry, Waltz explores interventions aimed at securing the borders between productive citizens, productive citizens in becoming, and citizens deemed not productive enough. Based on his scholarship of the twentieth-century philosopher Ludwig Wittgenstein, Nick Chown considers the possibility that neurotypical language games produce distinctions between autistic and non-autistic people from cultural biases in favor of neurotypicality. He elicits the implications of this on the medical-ethical framework of prevention and/or cure of autism debates. Virginia Bovell explores contrasting ethical approaches to the question of cure and prevention, including questions of identity, pressures for normalisation, and the kinds of suffering and harm that autistic people and their families actually experience.

In the second theme, **Neurodivergent wellbeing**, Robert Chapman and Alan Jurgens explore the social production of the category of autism and neurodiverse subjects through processes of cognitive othering, questioning neurotypical assumptions about human flourishing. In Chapman's case, this involves proposing a value-neutral model of disability in opposition to both social and medical models of disability. In Jurgens' case, it means utilising an enactive framework informed by the social model approach, from which social cognition and identity are stressed to be constituted via embodied and embedded social practices of institutions.

In the third theme, **Cross-neurotype communication**, Alyssa Hillary, Anna Stenning, and David Jackson-Perry with co-authors, focus on the specific ways in which we might challenge the social and cultural production of the normal subject, by dwelling in the neurodivergent social and cultural realm. This includes, in Alyssa Hillary's case, the possibilities of the application of cross-cultural communication principles to cross-neurotype communication. Stenning focuses on the ethical implications of neurodiversity, talking back against the characterisation of autism as deficits in empathy, presenting autistic life-writing as a source of understanding neurodivergent morality. Jackson-Perry and co-authors challenge dominant constructions of sensory experience by focusing on the experiences of neurodivergent people. The chapter also serves as an intervention in methodological discussions around neurodiverse collective knowledge production.

In the fourth theme, **Neurodiversity at work**,[3] Nicola Martin, Hanna Bertilsdotter Rosqvist with co-authors, and Matthew K. Belmonte, discuss further the contexts and conditions for neurodiverse knowledge production, among them barriers in academia and impact of cognitive normative ableist assumptions on the 'autistic loner' rather than the 'collective' in the case of autistic togetherness in knowledge production. They stress the impact of social and institutional contexts in expanding the potential for more inclusive research processes.

In the final theme, **Challenging brain-bound cognition**, Inês Hipólito and co-authors challenge the prevalent brain-bound picture of social cognition that promotes the view that the social cognitive profiles of autistic individuals are due to an underlying deficiencies, rather encouraging the adoption of alternative ways of understanding the social cognitive styles of the general population and autistic individuals that are not brain-bound. We believe this marks the way forward for a future neurodiversity studies that is able to loop between the different domains in which the term is used, rather than contribute to the 'othering' that has produced the marginalisation of neurodivergent people.

The book ends with a final section including three short chapters that aim to be looking forward at **the future of neurodiversity studies**, and our brief conclusion. In this way we offer a tentative roadmap for neurodiversity studies, where we put forward our own ideas as to some of the lines on which neurodiversity studies might develop.

Acknowledgements

With many thanks to Abby Ashley and Julia Bahner for the comments on a draft of this introduction.

Notes

1 According to Nick Walker (op. cit.) 'the terms neurodivergent and neurodivergence were coined by Kassiane Asasumasu, a multiply neurodivergent neurodiversity activist' (n.p.).
2 Although see Chapman's comments in Chapter 14 on the difficulties in describing impairments without reflecting the ideological assumptions of an ableist society.
3 The title of the theme 'Neurodiversity at work' involves word play which we know some autistic people thrive on but others have difficulty with. There is a double meaning here in that the theme refers to both neurodiversity in the workplace and practical applications of neurodiversity.

References

Baron-Cohen, S. (2015). Neuroethics of neurodiversity. In J. Clausen & N. Levy (Eds.). *Handbook of neuroethics* (pp. 1757–1763). Dordrecht: Springer.

Blume, H. (1997). Autism and the internet, or, It's the wiring, stupid. URL: (Retrieved December 2019) http://web.mit.edu/comm-forum/legacy/papers/blume.html

Broderick, A. A., & Ne'Eman, A. (2008). Autism as metaphor: Narrative and counter-narrative. *International Journal of Inclusive Education*, 12(5–6), 459–476.

Bumiller, K. (2008). Quirky citizens: Autism, gender, and reimagining disability. *Signs*, 33(4), 967–991.

Chamak, B. (2010). Autism, disability and social movements. *Alter*, 4(2), 103–115.

Ellsworth, E. (1989). Why doesn't this feel empowering? Working through the repressive myths of critical pedagogy. *Harvard Educational Review*, 59(3), 297–324.

Fenton, A. (2007). Autism, neurodiversity and equality beyond the 'normal'. *Journal of Ethics in Mental Health*, 2(2), 1–6.

Garland-Thomson, R. (1997). *Extraordinary bodies: Figuring physical disability in American culture and literature*. New York: Columbia University Press.

Hughes, J. M. (2015). *Changing conversations around autism: A critical, action implicative discourse analysis of US neurodiversity advocacy online* (Unpublished doctoral thesis). University of Colorado Boulder.

Jaarsma, P., & Welin, S. (2012). Autism as a natural human variation: Reflections on the claims of the neurodiversity movement. *Health Care Analysis*, 20(1), 20–30.

Kafer. A. (2013). *Feminist, Queer, Crip*. Bloomington, IN: Indiana University Press.

Kapp, S. K., Gillespie-Lynch, K., Sherman, L. E., & Hutman, T. (2013). Deficit, difference, or both? Autism and neurodiversity. *Developmental psychology*, 49(1), 59–71.

McRuer, R. (2006). *Crip theory: Cultural signs of queerness and disability*. New York: New York University Press.

Milton, D. (2014a). Autistic expertise: a critical reflection on the production of knowledge in autism studies. *Autism*, 18(7), 794–802.

Milton, D. (2014b). Embodied sociality and the conditioned relativism of dispositional diversity. *Autonomy, the Critical Journal of Interdisciplinary Autism Studies*, 1(3), 1–7.

Nicolaidis, C. (2012). What can physicians learn from the neurodiversity movement? *Virtual Mentor*, 14(6), 503–510.

Oliver, M. (1996). *Understanding disability: From theory to practice*. Basingstoke: Macmillan.

Ong, P. (2014). Neurodiversity and the future of autism. URL: (Retrieved December 2019) https://priceschool.usc.edu/wp-content/uploads/2013/07/Ong_Neurodiversity.pdf

Orsini, M. (2009). Contesting the autistic subject: Biological citizenship and the autism/autistic movement. In S. J. Murray & D. Holmes (Eds.), *Critical interventions in the ethics of healthcare: Challenging the principle of autonomy in bioethics* (pp. 115–130). Farnham: Ashgate.

Orsini, M. (2012). Autism, neurodiversity and the welfare state: The challenges of accommodating neurological difference. *Canadian Journal of Political Science*, 45(4), 805–827.

Perry, A. (2011). Neuropluralism. *Journal of Ethics in Mental Health*, 6, 1–4.

Perry, A. (2012). Autism beyond pediatrics: why bioethicists ought to rethink consent in light of chronicity and genetic identity. *Bioethics*, 26(5), 236–241.

Perszyk, D. (2013). Neurotypical. In F. R. Volkmar (Ed.), *Encyclopedia of autism spectrum disorders*. Springer: New York.

Runswick-Cole, K. (2014). 'Us' and 'them': The limits and possibilities of a 'politics of neurodiversity' in neoliberal times. *Disability and Society*, 29(7), 1117–1129.

Singer, J. (1999). Why can't you be normal for once in your life? From a 'problem with no name' to the emergence of a new category of difference. In M. Corker & S. French (Eds.), *Disability discourse* (pp. 59–67). Buckingham: Open UP.

Walker. N. (2014). Neurodiversity: Some basic terms and definitions. URL: (Retrieved January 2020) https://neurocosmopolitanism.com/neurodiversity-some-basic-terms-definitions/

Walker. N. (2015). Neuroqueer: An introduction. URL: (Retrieved January 2020) https://neurocosmopolitanism.com/neuroqueer-an-introduction/

Ward, R., & Price, E. (2016) Reconceptualising dementia towards a politics of senility. In S. Westwood & E. Price (Eds.), *Lesbian, gay, bisexual and trans* individuals living with dementia: concepts, practice and rights* (pp. 65–78). London: Routledge.

Yergeau, M. (2018). *Authoring autism: On rhetoric and neurological queerness*. Durham, NC: Duke University Press.

Part I

Curing neurodivergence/ eugenics

Chapter 1

The production of the 'normal' child

Neurodiversity and the commodification of parenting

Mitzi Waltz

The 'normal' child has only existed for about 100 years (Aries, 1962; Burman, 2007).

Two events converged to provide the tools for defining normalcy, and through that act, for defining its opposites: the advent of compulsory schooling, and the mass conscription of young males as cannon fodder for World War I. With some delays and regional variations, similar processes can be observed around the world as concepts of normalcy spread from the US and Europe, sometimes as part of colonial practices, sometimes as part of the spread of economic discourses (whether capitalist, fascist, socialist, or communist) predicated on appeals to modernity.

The first of these processes precipitated a clash between psychiatrists and psychologists in France over who could decide what happened to children who seemed to need help in primary school once every child was required to attend. Alfred Binet, a psychologist, pushed for including more children with different learning profiles in schools than in asylums. To locate and address their problems, Binet developed the first test of intelligence in 1905, the Binet–Simon test, later Americanised and further expanded as the Stanford–Binet test (Nicolas, Andrieu, Croizet, Sanitoso, & Burman, 2013). However, Binet's actions served as much to create an expanded professional role for the burgeoning profession of (educational) psychology as to protect children from being educationally excluded and/or institutionalised.

The second process revealed a high percentage of recruits who deviated from military expectations, or who later required treatment for 'shell shock' (post-traumatic stress disorder). This necessitated tools to formalise expectations by defining the parameters of 'normalcy': intelligence tests for adults, personality inventories, and so on. As Capshew (1999, p. 143) notes, 'personnel work, ranging from the initial selection of soldiers to the rehabilitation of combat casualties, was at the centre of psychologists' wartime efforts.' This focus was maintained after the war, both in the military's concept of modern weapons as 'man-machine systems' (ibid., p. 144) and in the burgeoning field of industrial psychology (van de Water, 1997).

Veterans of military psychology became the leaders of post-war applied psychology in the US (Capshew, 1999), including both industrial psychology and efforts aimed at individuals, families, and communities. These experts built on their wartime and post-war experience by founding the university programmes that cemented psychology as a graduate profession and established the norms it used to justify labelling and treatment. They led the professional organisations and edited the journals; many of their names feature prominently in early efforts to create all kinds of diagnostic and treatment guidelines.

But underneath both of these developments it was the mass unrest that marked the Industrial Revolution and the boom–bust cycles that drive capitalism that gave the figure of the 'normal' child its central place in medicine, social work, education, and parenting. Binet himself used the language of the 'social threat' posed by children who were excluded from schooling, positing a future of either being a social burden or part of the criminal class (Nicolas et al., op. cit.). The personality inventories that first appeared in the US and Europe during World War I to keep soldiers who might fall apart on duty away from heavy weaponry were soon thereafter put into service to identify the 'maladjusted' worker who might be tempted to join a union or subvert corporate goals (Gibby & Zickar, 2008). From their earliest years these tests included questions that could have been designed to weed out people on the autism spectrum, had autism then existed as a diagnostic category – for example, 'Do you get tired of people easily?' (ibid.) Psychologists advising industry suggested that problems with workplace agitation could be solved by getting rid of 'deviant' personalities in the workplace: Doncaster Humm (1943, cited in Gibby & Zickar, 2008, p. 167) suggested that the suspiciously exact figure of 80 per cent of employees causing problems in the workplace had personalities with a 'quirk or unusual feature'. Through such pronouncements, a diversity of neurological types became something to be feared, avoided, and potentially medicalised.

Humm's contribution to the burgeoning science of classifying 'deviant' personality types was the Humm–Wadsworth Temperament Scale (HWTS), which was developed to bring personality testing to a new market, the American workplace. Shaken by incidents of workplace violence, such as the 1934 murder of a supervisor by an employee (Hemsath, 1939), large employers flocked to add such tests to their pre-employment screening processes. The HWTS was heavily marketed to employers, and eventually became the inspiration for the personality inventories used most commonly today: the Meyers–Briggs Type Indicator and the Minnesota Multiphasic Personality Inventory (Lussier, 2018).

Humm based his 318-question survey on the work of noted eugenics advocate Aaron Rosanoff, who claimed that human personality traits could be classified as hysteroid, manic, depressive, autistic, paranoid, or epileptoid (1921). According to Rosanoff, these traits exist in all people to some degree, so the key factor in achieving normalcy was the degree of self-mastery found in an individual: the ability to damp down expression of these deviant impulses (ibid.).

Rosanoff's categorisation system was based on such dubious pseudoscientific work as word-association exercises with hundreds of subjects ('normal' students and workers, 'abnormal' state hospital patients), with an implicit assumption that what was most typical was also most desirable, and also no attention to issues like educational level or English proficiency (Kent & Rosanoff, 1910). In 1921, he was one of the first to describe an 'autistic personality type' (citing the definitions of Emil Kraepelin and August Hoch). At this point, 'autism' was typically used as a descriptor of specific avoidance behaviours in persons with schizophrenia, following the lead of Eugen Bleuler, but in this article Rosanoff pioneered by additionally identifying it with behaviours seen in persons functional enough to be in employment as well as with the diagnosis of dementia praecox (schizophrenia) (ibid.) An examination of early twentieth-century records substantiates that many individuals who would today be seen as having autism were given this label in the pre-Kanner years (Waltz & Shattock, 2004).

Rosanoff's assumption was that a stable norm existed, and any personality type that deviated from that norm was to a lesser or greater degree pathological. Crucially, his descriptions of the six 'abnormal' types all included information about how these issues presented during childhood. 'Normal' personalities, on the other hand, were characterised by 'inhibition, emotional control, a superior durability of mind, rational balance and nervous stability' (Rosanoff, 1921, p. 422). While he went on to note that some degree of 'abnormal' traits may be necessary for greatness (for example, 'How much in science and other fields, in which great concentration of mental energy on special tasks is required, is due to the inclination, peculiar to autistic personality, to exclude every diverting influence, every extraneous interest?' [p. 424], an early variation on the 'we're all a little bit autistic' discourse that will be familiar to modern scholars in critical autism studies), at the same time he valorised typicality and self-control. In his view, a person whose word-association responses were more than 50 per cent original rather than typical was likely in need of psychiatric care and control, and certainly not a good choice as an employee (Kent & Rosanoff, 1910).

And who was primarily responsible for children, who Rosanoff acknowledged as displaying and then (usually) growing out of traits that would be seen as problematic in adults? Parents, of course. The military and employers could only recognise the constellation of personality types that presented for entry, whereas parents were quickly positioned as controlling the efficacy of the production line.

The child as product

Peter Stearns maps the trajectory of constructing the child as particularly vulnerable and in need of scientifically based guidance especially well (Stearns, 2004). He points out key developments like the rise of magazines aimed at anxious parents in the 1920s, in response to the growing discourse about parental culpability for maladjustment and mental ill health, and the ever-changing list of things to be worried about that emerged over the ensuing decades. Stearns also places these developments squarely within the framework of rapid, challenging socioeconomic change: change that required the production of new, but still standardised, kinds of people.

Modern industry (like modern warfare) demanded adults who could be slotted into industrial processes as a standardised component. So, while early forms of testing for mental health difficulties, brain injury, and intellectual ability had been developed in the nineteenth century, their lack of standardisation was a stumbling block that prevented widespread use (Bondy, 1974). Once this hurdle had been overcome, however, tests were still often employed much too late to serve the military–industrial complex effectively. Thought among professionals soon quite logically turned to the issue of prevention: not merely recognising and excluding the abnormal adult but preventing abnormality in adults by intervening in childhood.

And so, the 'normal' child was constructed as an aspirational production goal for parents. This was done through a variety of processes, each offering services and products for its own target group of parents. The Child Guidance movement, for example, was a concerted effort to bring an end to the twin problems of juvenile delinquency and (future) workplace maladjustment. It can be seen as the child-focused variant of the Mental Hygiene field, which sought to prevent mental ill health, drug and alcohol abuse, and social unrest in adults through application of psychology and psychiatry. Despite its name, the focus of the Child Guidance movement was actually mothers rather than children.

Parental concerns were increasingly shaped by developments in psychology and psychiatry during this period. First the Child Guidance movement brought psychiatry from the private offices of elite practitioners to inner-city clinics (Jones, 1999). Alongside these 'child-saving' activities, which tended to be aimed squarely at working-class and poor families, came an army of experts who marketed their wares to middle-class mothers. Proponents like Ernest and Gladys Groves stated that most ordinary mothering was ''pathological': it did not produce children who became the kinds of adults who were in demand, and so a radical rethink was needed, with experts like the Groveses advising mothers on how best to carry out their duties (Waltz, 2016).

Compulsory schooling provided fertile ground for theorists, researchers, and practitioners to explore both familial and non-familial production strategies – but it also acted in turn to further constrain the figure of childhood normalcy. As Stearns writes, 'the very successes achieved in improving children's lives led to an escalation in what came to be seen as the minimal standard for children's well-being' (Stearns, 2004, p. 2). These standards were more frequently than not laid at the doorstep of parents, who were expected to provide guidance in both academic and social performance.

Of course, this push for standardisation was also closely entwined with eugenics, a pseudoscience then in ascendance throughout the Western world and beyond. So, while on the one hand children were assessed in Child Guidance clinics and mothers were given 'scientific' child-rearing advice, some forms of aberrance were instead addressed with removal to institutional care, forced sterilisation, or euthanasia (and not only in Nazi Germany, where this process predated the Holocaust to come). The approach taken often depended more on which service was seen as appropriate for the family in question, and there was a clear class dimension to this, with middle- and upper-class families somewhat more likely to receive clinical help, while incarceration and/or eugenic solutions (typically targeted birth control programmes, segregation via institutionalisation, and sterilisation) were more often recommended for lower-class families and those from ethnic minorities (American Eugenics Society, 1936; Mazumdar, 1992).

Indeed, by the 1930s eugenicists were publicly declaring that about 10 per cent of the population was responsible for most of its social problems, and publishing lists of exactly which behaviours or characteristics could be used to recognise and segregate this group. While intelligence was one such criterion, others included epilepsy, unemployment, poverty, casual labour, and contact with the criminal justice system (Mallett, 1931). In other words, the ideological drivers behind eugenic policies saw not just one form of atypical mind, that of the intellectually disabled person, as problematic. Instead, it was believed that a variety of different atypical minds could and should be identified as producing social pathology, and that a project of identifying and addressing all of these would be far more effective than targeting only the 'feeble-minded'. These forms of neurological difference were often seen as separate but linked, with varied forms of genetic interaction producing varied forms of aberrance. These were, in turn, medicalised whenever possible.

The pathologisation of neurodiversity and the rise of autism

And so it was that only in the heyday of the Child Guidance movement in the US and Europe could a diagnosis such as autism become possible. Accurate labelling of 'problem children' was seen as the first step to correcting the various family problems that were believed to underlie each form of pathology.

> The nervous child, the delicate child, the enuretic child, the neuropathic child, the maladjusted child, the difficult child, the neurotic child, the over-sensitive child, the unstable child and the solitary child, all emerged as a new way of seeing a potentially hazardous childhood.
>
> (Armstrong, in Scambler, 2005, p. 237)

Once a category could be labelled and defined, specific recommendations were generated for mothers. Diagnosis, assessment, and services were typically delivered in outpatient settings, augmented by home visits, but the threat of child removal by social services and relegation to institutions was always there for those families whose parenting came under the harsh glare of the Child Guidance microscope.

Leo Kanner's development of the autistic child as a specific category is best seen as part of this project. Most of the patients profiled in his 1943 article, which defined childhood autism for the first time in English, were recruited from a Child Guidance clinic, the Harriet Lane Home. Assessments performed by Child Guidance professionals were often the basis for his own reports (Waltz, 2013). Kanner's assumption that parents were to blame for this form of difference was clear in his case studies, though later refuted to some extent by their author, and this orientation coloured the advice given for many decades thereafter (ibid.; Silberman, 2015). And Kanner was not alone in his connection to this movement: many other pioneers in autism theory and practice also worked within the Child Guidance context, from John Bowlby in England, who started his career in it and later led children's services at the Tavistock Institute, to Bruno Bettelheim, who was director of the Sonia Shankman Orthogenic School at the University of Chicago. The Orthogenic School had been set up as a kind of demonstration project to prove that Child Guidance principles could be successfully applied to a broad spectrum of 'disturbed' children (very few of whom met the criteria for autism) (op. cit.; Pollak, 1998).

But it was not only children who were seen as thinking or behaving differently who were pathologised by this process; mothers (and to a lesser extent, fathers) were affected as well. The rise of 'scientific' motherhood impacted mothers of children seen as troubled or troublesome most directly, but it also informed a growing – and closely connected – childrearing advice industry. The 'normal child' was increasingly surrounded by surveillance of physical growth, behaviour, language development, and academic achievement. New ways of being a 'deviant child' arose, such as the medicalised diagnosis of dyslexia, but parents also had to cope with demands to teach all of their children to regulate emotional expression, behaviour, desires, and achievement. As Stearns notes, emotions like fear, jealousy, and anger that were once seen as relatively normal in children were themselves pathologised as the process gained pace – now not only should they not be expressed, they should not even be felt. 'Adulthood itself became more complex in many ways, which meant that children must adjust accordingly.... Emotional health itself was redefined, becoming harder to achieve' (Stearns, 2004, p. 49).

It is often said that the most powerful motivators in advertising are fear and jealousy, and this was certainly the case for the childrearing advice industry that arrived to help parents in the developed world navigate this confusing terrain. Indeed, it is interesting that two of the very emotions that parents were being asked to eliminate were deployed to motivate them. The figure of the 'aberrant child' could be used as an object of fear to encourage parents to follow prescribed patterns of childrearing, even when they did not seem to fit the child they actually had in front of them, while the figure of the 'normal child' could be deployed as an aspirational goal, always just out of reach.

To ensure that as many adults as possible were available for work, rationalised models of parenting were introduced to reduce the labour 'wasted' on resistantly neurodivergent children and to lessen the impact of 'incompetent' parenting, with the state and charitable institutions poised to take over as mass corporate parents. As the child abuse scandals currently roiling every part of the planet have revealed, the state, the Church, and charities all proved to be terrible parent-substitutes. However, their parenting roles were cemented in place for decades by the pathologisation of neurodiversity in children, and in the policing of parenting.

And as the decades marched on, the definitions used to diagnose and address neurological difference expanded to capture far more children than was the case even in the heyday of eugenics, with its talk of a problematic 10 per cent.

The current state of regulated normalcy in childhood

One hundred years later, while large-scale institutions for children are far less common, the family home itself has become the centre of rationalised, medical-ised, preventative parenting. Child development is routinely measured against charts and schedules, and surveillance now begins even before birth, with ultra-sound imaging of the foetus and genetic testing. While biomedical notions are fre-quently referred to in current discourses of childhood aberrance, parent-blaming is never far away, because parental labour is the cheapest remedy and because it is so deeply ingrained in culture. As William Kessen has written:

> The tendency to assign personal responsibility for the successes and failures
> of development is an amalgam of the positivistic search for causes, of the
> older Western tradition of personal moral responsibility, and of the conviction
> that personal mastery and consequent personal responsibility are first among
> the goals of child rearing
>
> (1979, p. 819)

Standardised testing has moved beyond the measurement of intelligence and physical dexterity to become the cornerstone of education, and a key element of the pathway to work. Most school systems worldwide are characterised by a series of key stage exams, any of which can be used to identify and re-route children who deviate from the norm. A multiplicity of normal and abnormal personality types and ways of learning have been defined, but despite rhetoric about 'multiple intelligences' (Gardner, 2006) and diversity, only some of these are privileged.

Indeed, the current vogue for standardised parenting might as well come with an explicit analogy to Ford's assembly line. To use just one example: 'Triple P takes the guesswork out of parenting!' trumpets the website for an approach that bills itself as 'evidence-based' and 'for every parent' … 'shown to work across cultures, socio-economic groups and in many different family struc-tures' (Triple P International, 2019). This particular approach, like many of its competitors, is a commercial product: to use the 'brand name', practitioners must attend costly trainings and gain the permission of its progenitors, Triple P International. It uses new names for what Rosanoff so many decades ago called 'self-mastery': today's buzzwords are 'self-sufficiency', 'emotional resilience', and 'self-regulation' (for example, Newland, 2014). The programme has been designed to offer the illusion of individualisation and choice, as expected by the modern consumer – parents can attend individual sessions or in groups, and variants for parents of teens or children with 'special needs' are available, but all offer the same essential content. There is a version for the parent who is merely anxious, another for the parent who is definitely experiencing challenges, but for both the strategies are the same. Each iteration of the programme will require provision of workbooks and other materials provided by Triple P International. And while Triple P and similar parenting programmes can be chosen voluntarily, many families are ordered to attend by social service agencies or courts, with specific goals set that they and their child must achieve to avoid more harsh directives.

These developments reflect societal changes that are spreading far from their Western origins. Workers today are required to exercise not just emotional control, the issue that so concerned the Child Guidance movement in the early twentieth century, but also to have a high capacity for emotional performance and for the management of the emotions of others (Ehrenreich, 2010). This is quite a shift in expectations, evidencing ever more restricted norms of appropriate social behaviour. It presents a figure of the 'normal child' that is gendered, racialised, and set within a specific set of socioeconomic class relations: for example, it is a figure with the 'emotional control' once positioned as a male trait but also the 'emotional intelligence' previously ascribed more often to women; a figure able to perform emotion publicly in a way that is dictated by Western cultural elites, quite possibly in conflict with other existing cultural norms. And while the recent emphasis on emotional intelligence can be seen as emphasising collective over individual behaviour, this is actually an illusion: the entire point of such performance is manipulation of others to achieve personal or employers' goals (ibid.; Hochschild, 1983; Bates, 2015).

Here is where autism and other forms of neurodiversity have been rolled out as spectres to be avoided or eliminated, and here is where we locate the technologies that are brought to bear when deviance is feared or suspected. The child who is more active than his peers becomes 'hyperactive', the child who struggles at school becomes 'dyslexic', the child who fails to conform to gendered behaviour norms has 'gender dysphoria'. For each form of aberrance, a set (or often multiple, competing sets) of educational and medical technologies is deployed to pull a specific type back towards the norm. While labels can theoretically be used to provide support that allows each child to flourish, the desire for standardised outcomes tends to guide the form and goals of support in education systems as they currently exist.

The language of developmental surveillance itself wants modern parents to be alert and afraid: it is rife with talk of 'red flags' and 'warning signs'. And as the definition of autism now includes a more significant number of children who do not have intellectual disabilities than of those who do, parents are encouraged to be ever watchful of the child who may have trouble making friends (or who simply prefers their own company), the child whose language development is 'too early', the child who struggles in social learning environments.

And despite the medicalisation of autism and other forms of neurodiversity, as noted previously, the figure of parental causation is never really outside the frame. Production of the 'normal child' begins at home, especially once professionals start making noises about 'red flags'. Meeting expectations requires significant physical and emotional labour on the part of parents, and ensuring that the job is properly done requires both official and self-surveillance, especially of parents' emotional labour.

For children whose neurodiversity is not easily made to disappear, the consequences can be all-encompassing. Imagine, if you will, the experience of up to 40 hours per week of behavioural training, delivered in an endless series of tasks that demand an immediate, correct response. Or the experience of growing up in a family home where your parents are supervised to deliver a 24-hour 'therapeutic environment'. Imagine a residential centre where every form of speech or behaviour deemed aberrant by staff results in an electric shock. Imagine a life spent measuring oneself against a moving target, with economic and emotional consequences at every juncture.

It is potentially a horror on an individual level, but is also a highly profitable industry. Essentially, these technologies, whether they are medical or quasi-educational, reduce the 'normal child' to a standardised product, one whose parameters are regularly redesigned to fit the needs of the state and the labour market.

Producing advantage and disadvantage

It must also be said that the 'child-saving' technologies marketed to Western and other privileged parents are not accessible to the majority of families worldwide. This creates a paradox where 'normalcy' actually comes to describe an ideal type rather than the literal norm, with all its potential for divergence and messy boundaries. A new hierarchy of über-normalcy has arisen, where only the most well-heeled can produce a child who fits the mould perfectly, with exclusion of others easily blamed on personal factors rather than lack of opportunity, economic inequality, race, or culture. On one side of this equation lies the 'Tiger Mom' who pushes her neurotypical child towards perfection, with endless rounds of cram classes, music lessons, and enrichment activities. On the other lies the mother of humble means who is concerned by receiving warnings about signs of developmental disability, but who cannot afford the interventions marketed to wealthy parents. This mother is the perfect mark for con artists peddling fake, cheap, and potentially harmful 'treatments' such as bleach enemas, which are increasingly marketed to carers in developing countries and to working-class and immigrant carers in the West (Gorski, 2019; Meershoek, 2019). She is a likely target for 'cost-effective' brand-name parenting programmes, parenting apps, and parent-shaming.

The ever-smaller definition of 'normal' can therefore also be seen to represent a way for the wealthy to maintain their economic advantage. Even as well-to-do families use dubious diagnoses of dyslexia or ADD to gain advantage for their off-spring on university exams (Medina, Benner, & Taylor, 2019), the same diagnoses are used by institutions that face poor and working-class youth as a springboard to the B-track in life or, worse yet, as a gateway to the school-to-prison pipeline. As Emily Boyk notes,

> When disability is combined with discourses of race and violence, disabled students morph from burdens to active threats…. Importantly, the discourses that fuel this devaluation are operationalised through policies like the Individuals with Disabilities Education Act that, while theoretically empowering disabled students, in many ways continue to reproduce oppressive understandings of disability.
>
> (Boyk, 2018, p. 5)

The danger that this oppression will include new forms of eugenics is never far away: for example, see the pseudoscientific musings of Dominic Cummings (2013) who was as of this writing the lead advisor to the British prime minister. As Vizcarrondo (2014) has written, the

> new eugenics proposes to create better opportunities for children through individual human enhancement and undesirable trait elimination…. This intervention results in the design of the child to the parents' expectations. The design of the child according to the parents' selection of traits may leave little room for consideration of the child's choices [with the end result that] the child may be viewed as a product.
>
> (pp. 240–242)

While Vizcarrondo points to 'parents' expectations', these expectations are produced through socioeconomic processes.

Having first produced inequalities and oppression, our economic system has adapted admirably to profit from these demonstrable failures. The production of the 'normal child' has become a highly segmented and specialised industry, and the latest development in this industry has been the commodification of deviance itself in additional profit centres: the autism industry, the ADD/ADHD industry, the dyslexia industry. Even when normalcy cannot be produced by the system, its failures can be monetised as students and clients.

To conclude, the policing of neurodiversity as a way of defining the 'normal child' can be described as perhaps the most obvious example of biopower in action: the application of control mechanisms to populations defined as deviant, based on appeals to science, in order to constrain the behaviour of both the majority and the rest (Foucault, 1978). Although the forms and targets of these controls have changed over time, their result has been to produce definitions of deviance that capture an ever-larger proportion of the human population, predicated on an increasingly restricted definition of the 'normal child' and the application of varied medical and quasi-educational technologies to bring neurodiverse outliers into compliance with this figure. This analysis is not intended to erase or question neurodiversity, but to draw attention to how observation of difference can be used to produce specific forms of societal relations rather than to support and celebrate individuality and varied forms of intelligence and behaviour.

References

American Eugenics Society. (1936). Memorandum by the American Eugenics Society: A eugenics program for the United States. *Eugenics Review, 27*(4), 321–326.

Aries, P. (1962). *Centuries of childhood: A social history of family life*. New York: Knopf.

Bates, A. (2015). The management of 'emotional labour' in the corporate re-imagining of primary education in England. *International Studies in Sociology of Education, 26*(1), 66–81.

Bondy, M. (1974). Psychiatric antecedents of psychological testing (before Binet). *Journal of the History of the Behavioral Sciences, 10*(2), 180–194.

Boyk, E. (2018). *No IDEA: Biopower and the school-to-prison pipeline for students with disabilities* (Honors Thesis Collection, Wellesley University). URL: (Retrieved August 2019) https://repository.wellesley.edu/thesiscollection/520

Burman, E. (2007). *Deconstructing developmental psychology*. London: Routledge.

Capshew, J. H. (1999). *Psychologists on the march: Science, practice, and rofessional identity in America, 1929–1969*. Cambridge: Cambridge University Press.

Cummings, D. (2013). Some thoughts on education and political priorities. *Guardian.* URL: (Retrieved December 2019) www.theguardian.com/politics/interactive/2013/oct/11/dominic-cummings-michael-gove-thoughts-education-pdf

Ehrenreich, B. (2010). *Smile or die: How positive thinking fooled America and the world.* London: Granta.

Foucault, M. (1978). *The history of sexuality, Part I: The will to knowledge*. New York: Pantheon Books.

Gardner, H. (2006). *Multiple intelligences: New horizons in theory and practice*. New York: Basic Books.

Gibby, R. E., & Zickar, M. J. (2008). A history of the early days of personality testing in American industry: An obsession with adjustment. *History of Psychology*, 11(3), 164–184.

Gorski, D. (2019). Bleaching away what ails you: The Genesis II Church is still selling Miracle Mineral Supplement as a cure-all. *Science-Based Medicine*, 22 April. URL (Retrieved August 2019): https://sciencebasedmedicine.org/miracle-mineral-supplement-as-a-cure-all/

Hemsath, M. E. (1939). Theory and practice of temperament testing. *Personnel Journal, 18*, 3–12.

Hochschild, A. (1983). *The managed heart: Commercialisation of human feeling*. Berkeley, CA: University of California Press.

Jones, K. W. (1999). *Taming the troublesome child: American families, child guidance, and the limits of psychiatric authority*. Cambridge, MA: Harvard University Press.

Kent, G. H., & Rosanoff, A. J. (1910). A study of association in insanity. *American Journal of Insanity, 67*(37–96), 317–390.

Kessen, W. (1979). The American child and other cultural inventions. *American Psychologist*, *34*(10), 815–820.

Lussier, K. (2018). Temperamental workers: Psychology, business, and the Humm-Wadsworth Temperament Scale in interwar America. *History of Psychology*, *21*(2), 79–99.

Mallet, B. (1931). The social problem group: The President's account of the Society's next task. *Eugenics Review*, *23*(3), 203–206.

Mazumdar, P. (1992). *Eugenics, human genetics and human failings: The Eugenics Society, its sources and its critics in Britain*. London: Routledge.

Medina, J., Benner, K., & Taylor, K. (2019). Actresses, business leaders and other wealthy parents charged in U.S. college entry fraud. *New York Times*, 12 March. URL: (Retrieved August 2019) www.nytimes.com/2019/03/12/us/college-admissions-cheating-scandal. html

Meershoek, P. (2019). Amsterdamse bisschop promoot dubieus 'medicijn'. *Het Parool*, 16 February. URL: (Retrieved August 2019) www.parool.nl/amsterdam/amsterdamse-bisschop-promoot-dubieus-medicijn~b38ea82a/

Newland, L. A. (2014). Supportive family contexts: Promoting child well-being and resilience. *Early Child Development and Care*, *184*(9–10), 1336–1346.

Nicolas, S., Andrieu, B., Croizet, J.-C., Sanitoso, R. B., & Burman, J. T. (2013). Sick? Or slow? On the origins of intelligence as a psychological object. *Intelligence*, *41*(5), 699–711.

Pollak, R. (1998). *The creation of Dr. B.: A biography of Bruno Bettelheim*. New York: Touchstone.

Rosanoff, A. J. (1921). A theory of personality based mainly on psychiatric experience. *American Journal of Psychiatry*, January, 417–436.

Scambler, G. (2005). *Medical sociology: Major themes in health and social welfare, Vol. 1: The nature of medical sociology*. London: Routledge.

Silberman, S. (2015). *NeuroTribes: The legacy of autism and how to think smarter about people who think differently*. New York: Avery.

Stearns, P. N. (2004) *Anxious parents: A history of modern childrearing in America*. New York: NYU Press.

Triple P International. (2019). *Triple P Parenting Program Corporate Site*. www.triplep.net/glo-en/home/ Queensland: Triple P International Pty Ltd.

Van de Water, T. J. (1997). Psychology's entrepreneurs and the marketing of industrial psychology. *Journal of Applied Psychology*, 82(4), 486–499.

Vizcarrondo, F. E. (2014). Human enhancement: The new eugenics. *The Linacre Quarterly*, *81*(3), 239–243.

Waltz, M. (2013). *Autism: A social and medical history*. London: Palgrave Macmillan.

Waltz, M., & Shattock, P. (2004). Autistic disorder in nineteenth-century London. *Autism*, *8*(1), 7–20.

Chapter 2

Language games used to construct autism as pathology

Nick Chown

Ludwig Wittgenstein counselled against the bewitchment of our intelligence by means of our language, by which he meant that we risk misunderstanding something as a result of failing to notice logical errors in language used to describe it. Some scholars believe that autistic thinking is more individualistic and less likely to be stuck in the rut of conventionality. So, an autistically neurodivergent perspective on language use is valuable in identifying the bewitchment of intelligence that concerned Wittgenstein. I also argue that a failure of neurotypical society to appreciate that societal language games are, by definition, *neurotypical* language games has adverse consequences for autistic people because of the inevitability of cultural biases favouring neurotypicality. The philosopher Sandy Grant has written that

> as long as there is language it will bewitch us, we will face the temptation to misunderstand. And there is no vantage point outside it. There is no escape from language-games then, but we can forge a kind of freedom from within them.[1]

Might it be possible for an autistic person to escape a neurotypical language game – and all language games *are* neurotypical – and observe it from an external vantage point?

Wittgenstein introduced the concept of language games. Various of his ideas – including the language game concept – are relevant to an understanding of autism. A language game is the language associated with a particular activity that gives the activity its meaning. For example, the job interview is an activity where language is used in special ways. When an interviewer asks an interviewee to talk about their weaknesses, both parties should know that the response has to demonstrate self-awareness on the part of the interviewee; to provide a detailed description and analysis of weak points would be to misunderstand this particular language game. It is my view that the term 'language game' does not do full justice to Wittgenstein's intention because it implies a sole focus on language rather than the social interaction of which language is a part (albeit a very important part). Szasz refers to the 'game-playing model of human behavior' (Szasz, 2010,

p. 250) and to the importance of 'rules'[2] in human social interactional game-playing. I believe this is what language games are about.

While neurotypical language relating to autism inevitably reflects neurotypical perspectives on autism, societal understandings of autism will benefit from autistic perspectives that reflect the lived experience of autism. For instance, many autistic people consider that the monotropism theory of autism – developed by neurodivergent scholars – describes what it is like to be autistic better than any other theory (Murray, Lesser, & Lawson, 2005). And the double empathy hypothesis (Milton, 2012) – which draws attention to the bi-directional nature of the difficulty autistic and non-autistic people often have understanding each other – was also developed by a neurodivergent scholar. In addition to the language game concept, I draw attention to Wittgenstein's counsel against bewitchment of our intelligence through misuse of language. On occasions, and perhaps due to the subtlety of language, we draw conclusions that appear sound but that on investigation are found to be illogical. For example, the concept of the broader autism phenotype – which is thoroughly embedded in medical understandings of autism – is based on the illogical assumption that a cluster of traits used to screen for autism, and that any human being may present with, are somehow 'autistic traits' indicative of a subclinical presentation of autism in the general population known as the broader autism phenotype (Chown, 2019). This chapter begins a Wittgensteinian analysis of aspects of societal language use to demonstrate the value of a neurodivergent perspective in the identification of researcher misunderstandings of aspects of autism with the potential to influence ethical consideration of research to cure/prevent autism adversely.

First, it will be demonstrated that a failure to appreciate that societal language games are *neurotypical* language games can have adverse consequences where autism is concerned because of the inevitability of cultural biases in favour of neurotypicality. Second, it will be demonstrated that misuse of language can give rise to false beliefs about autism that may become embedded as received opinion in language games. In the first situation, the value of 'missing' neurodivergent perspectives will be shown directly. In the second situation it is contended that more individualistic (and possibly also more logical) thinking styles in autism may enable identification by autistic scholars of language misuse, that might otherwise remain hidden, as the thought processes of autistic people are less likely be influenced by pre-existing conceptual frameworks.

A substantial amount of autism research and its associated funding and publicity is focused on genetics, neuroscience, and the search for a cure (Pellicano, Dinsmore, & Charman, 2014). Although there has been considerable discussion of ethical matters in the autism research literature, most of this discussion refers to what one might call 'micro' ethical subjects such as informed consent and anonymisation. These subjects are important but of no relevance to an investigation of the ethics of research to eradicate autism. This is because discussion of 'micro' ethical subjects presupposes that the research being undertaken is research that is ethically valid. I describe a fundamental issue, such as whether a particular type of research should be undertaken at all, as a 'macro' ethical subject. Researchers rarely, if ever, discuss their justification for undertaking their study. There has been very little discussion of the ethics of autism cure/prevention in the literature (Bovell, 2015). Virginia Bovell's work is one of only two thesis-length discussions of this subject. She notes that there has been very little attempt to define the terms 'cure' and 'prevention' in relation to autism. Pursuance of a cure for autism has been problematised on ethical grounds by only a limited number of scholars. For instance, Majia Holmer Nadesan has written of the 'latent dangers lurking in a geneticization of autism devoid of environmental mediation' as well as the 'potential for ... prenatal testing potentially ushering in a new eugenics' (Nadesan, 2013, p. 137).

Wittgenstein's view of moral justifications is summed up well in the following quotation:

> Nothing we can do can be defended absolutely and finally. But only by reference to something else that is not questioned. I.e. no reason can be given why you should act (or should have acted) *like this*, except that by doing so you bring about such and such a situation, which again has to be an aim you accept.'
>
> (Wittgenstein, 1948, p. 16, author's italics)

If, like me, you believe him to be correct that there are no categorical imperatives or deity-given moral compasses, and therefore no absolute and final justification for what one does (and doesn't do), you will also agree with me that those who advocate eradicating autism must accept its eradication as a justifiable aim per se. This is presumably because in their view it is a disorder, and disorders are, by definition, harmful, and thus at odds with living a good life. It seems that most of those who would eradicate autism if they could, undertake their research on the basis of an aim they accept as a 'given', or at least without being willing to be transparent about their justification. Pellicano and Stears (2011) tell us that scientists defend the spending of the vast majority of autism research funds on research into genetics and neuroscience on the basis that: (1) identifying children at risk for autism before they show signs of autism will enable much earlier intervention than is currently the case, and (2) there will be medical benefits to improve the health of autistic individuals. If, indeed, these are the main defences used to justify such research, they appear disingenuous. This is due to the apparent focus on benefiting autistics being in clear contrast to the emphasis on seeking a cure for autism – sometimes expressed as 'prevention' – of funding bodies such as the National Alliance for Autism Research, Cure Autism Now, and Autism Speaks. Bovell (2015, p. 49) writes that 'sometimes the purpose of [autism] investigations falls short of any kind of articulated explanation beyond a "knowledge for knowledge's sake" perspective' which holds that 'potential benefits are somehow self-evident' (ibid., p. 50). She concludes that much autism research is based on autism being a 'bad thing' and cure a 'good thing'.

Where scientists justify research with the potential (if not the specific aim) of eradicating autism or other categories of neurodivergence, on a simple belief in the importance of seeking a cure for diseases and mental disorders, the issue is that it is not at all clear that these categories *are* mental disorders or are *always* mental disorders. Many autistic self-advocates and others have put forward a case that autism is neurological difference coupled with societal oppression as understood by the social models. While the language game associated with the 'cure' of diseases and disorders is uniformly positive, as indeed it should be, the inclusion of a phenomenon within the diagnostic manuals giving legitimacy to the search for a cure, is a matter for both political and scientific debate (Kapp, 2019). This can lead to the inclusion of diagnoses in the manuals that are categorically *not* diseases or disorders, with all the adverse consequences of such bad decision-making. One only has to consider the situation regarding gays and lesbians to appreciate that inclusion of a so-called disorder in a diagnostic manual can be problematic. Certain sexual orientations were included in the Diagnostic and Statistical Manual of Mental Disorders until as recently as 1987 and it was another three years before the World Health Organization removed the same orientations from their International Classification of Diseases (ICD-10). Debates about sexuality then shifted from psychiatry into the moral and political spheres as institutions could no longer justify discrimination against gay and lesbian people on the basis of (supposedly) scientific arguments used to pathologise them. Drescher (2015, p. 572) writes that

> Most importantly, in medicine, psychiatry, and other mental health professions, removing the diagnosis ['homosexuality'] from the DSM led to an important shift from asking questions about 'what causes homosexuality?' and 'how can we treat it?' to focusing instead on the health and mental health needs of LGBT patient populations.

Neurodiversity advocates would like to see similar developments in relation to autism. While many advocates support the search for a cure for conditions co-occurring with autism (co-morbidities) such as anxiety, gastrointestinal disorders, sleep disorders, and epilepsy (ibid.), that is because – unlike autism itself – they do not regard these as being core to the very nature of their being.

There have only been a limited number of investigations into the ethics of eradicating autism to date (e.g. Anderson, 2013; Barnbaum, 2008; Barnes & McCabe, 2011; Bovell, 2015; Chapman, 2019; Pellicano & Stears, 2011; Walsh, 2010). These authors all take an anti-discriminatory, anti-eradication stance except for Barnes and McCabe,[3] whose work is an investigation of the issue of choice (whether a cure should be made available for those who want one), and Barnbaum who writes that there is 'something intrinsically limiting in an autistic life' and appears to support the eradication of autism (Barnbaum, 2008, p. 154).[4] Anderson considers autism to be a valid identity and possibly even to have given rise to a culture. Walsh has challenged those who would prevent disability coming into the world, pointing out that preventing Asperger's would of necessity mean that the exceptional abilities associated with it would be lost to society. Liz Pellicano and Marc Stears set out an ethical objection to cure and prevention of autism but, importantly, one that only applies in the context of *living individuals*. Robert Chapman challenges the assumption underlying the dominant view of autism that it is inherently at odds with the ability to lead a good life. He concludes that there is no 'decisive reason to think that being autistic, in and of itself, is at odds with either thriving or personhood' (Chapman, 2018, p. 1). Bovell considers that research to cure/prevent autism is ethically indefensible. After unpacking the issues surrounding the ethics of curing/preventing autism she concludes that 'reference to prevention and/or cure as a desirable *general* goal[5] is neither clinically/ scientifically coherent nor morally legitimate' (Bovell, 2015, p. 364, author's italics). Her point that 'To talk in approving terms about prevention and cure implies that a world where there are no more autistic people would be a better world' (ibid., p. 364) is the thinking that lay behind the call for scientists engaged in research to cure and/or prevent autism to justify the ethical validity of their work (Chown & Leatherland, 2018).[6]

As already stated, the fundamental point here is the vexed question as to what autism is; is it a mental disorder or disease or a natural human difference? Bovell calls this the 'analogy challenge' as both positive analogies and negative analogies have been drawn in relation to autism. There is no definitive answer to this question as yet. Many autistic scholars believe that no researcher should ever assume that it is appropriate to seek to destroy any aspect of humanity without societal acceptance of the justification for their work, an acceptance that must be based on the most thorough of investigations and debates because the very survival of a category of people depends upon it. My aim here is to indicate how a Wittgensteinian grammatical perspective can uncover hidden instances of language bewitchment of relevance to the undertaking of autism cure/prevention research. The relevance arises from the risk of misleading our attempts to understand what autism is. This can lead to situations where issues become separated from concerns about their morality. Baumann refers to such separation as 'adiaphorisation' which he defines as 'stratagems of placing, intentionally or by default, certain acts and/or omitted acts regarding certain categories of humans *outside* the moral-immoral axis – that is, outside the "universe of moral obligations" and outside the realm of phenomena subject to moral evaluation' (Bauman & Donskis, 2013, p. 40, author's italics). He says that exemption of adiaphoric acts from ethical consideration due to social consent enables those acts to be committed without those involved facing any moral stigma or needing to worry their consciences about them. Scholars' failure to discuss the ethics of their research, and society's failure to call scholars to account for their failure, is adiaphorisation. The ethics of autism research must be brought into the 'universe of moral obligations'. Wittgenstein argued that certain aspects of language use can bewitch our intelligence. This chapter discusses examples of language misuse giving rise to false beliefs about autism.

Wittgensteinian grammatical investigation of autism language

By taking steps to avoid language games 'bewitching our intelligence' we will be in a better position to see concepts for what they really are, not what they appear to be when language clouds the understanding. A Wittgensteinian grammatical investigation[7] involves an exploration of a language game and the rules governing it, not an investigation of language structure. His primary focus was on the confusions that misuse of words can cause. This chapter discusses examples of language confusion that impact upon debates relating to the ethics of autism because they give rise to false beliefs about autism: neurotypical[8] language games;[9] illogical language moves; and confusing language.

Neurotypical language games

Milton's double empathy hypothesis argues that communication difficulties between neurotypical and autistic people are bi-directional in nature. Hughes (2019, personal communication) refers to such difficulties as reciprocal misunderstandings. If arguing that misunderstandings on the part of autistic people arise from a cognitive defect associated with autism, it could be argued on the basis of double empathy that the difficulties neurotypical people have understanding autistic people is due to a cognitive defect in neurotypicality. Alternatively, the difficulties autistic and neurotypical people have in communicating with each other could be due to society's language games being neurotypical language games based on neurotypical understandings of autism.

First, let us consider an issue arising from a medical language game taken from Bovell (2015, p. 280). She writes that 'engaging with the community of people who are most affected and able to reflect on [intervention practices] is likely to be essential, given the sorry history of autism having been drastically misunderstood by "outsiders" in the past'. Examples of misunderstandings include autism being caused by poor mothering (Kanner/Bettelheim); autism involving social isolation (Kanner); autism only affecting children (Kanner); autistic people being intellectually disabled; autistic individuals being unable to feel or express emotions; all autistic people lacking empathy and/or theory of mind. There are many more myths and misunderstandings, and they have probably all been perpetuated by non-autistic scholars who have just as much difficulty empathising with autistic people as vice versa[10] (Chown 2014; Milton, 2012). Milton and Bracher (2013) argue that the absence of autistic voices from work to generate knowledge about autism results in both epistemological and ethical problems as non-autistic people cannot have lived experience of being autistic. Unless and until medical language games of autism are allowed to develop with contributions from autistic scholars, they will remain prone to perpetuating misunderstandings about autism that impact theory and practice.

Let us now consider a cultural language game. Sarah Pripas-Kapit (2020, p. 25) writes that while

> authors such as Temple Grandin and Donna Williams introduced mainstream audiences to the concept of autistic people narrating their own experiences, their works still relied on ableist ideas about autism promoted by non-autistic scientific "experts" and parents. They positioned autism as a tragedy.

She points out that Sinclair (1993), who had a thorough understanding of parental perspectives on autism, challenged the assumption that autism is always tragic and that parental grief for the 'loss' of the expected child is the inevitable result of autism. The 'autism as tragedy' trope is an example of a cultural perspective on autism inextricably linked with neurotypicality,[11] and with which most autistic self-advocates would disagree. Parents of autistic children have contributed to the development of the neurodiversity movement, and some autistic individuals agree with the tragedy trope, so it is wrong to speak of necessarily opposed neurotypical and autistic attitudes. However, with cultural attitudes towards autism having developed in a neurotypical society where cure and prevention discourses are prominent, the language games of autism have inevitably developed in accordance with neurotypical society's cultural biases. Autistic self-advocates and the developing autistic online culture are effecting some change to this situation but unless and until autistic viewpoints are accepted as valid this cultural bias will continue.[12]

Illogical language moves

It is argued that those who believe in the existence of a broader autism phenotype (BAP) are led astray by misuse of language (Chown, 2019). Human traits indicative of autism are included in a screening cluster for a good reason – that there is a strong indication of autism if an individual has the cluster traits – but scholars then generally make the unjustified leap into thinking that having some of the cluster traits implies the existence of a broader autism phenotype of individuals who do not justify a diagnosis of autism but have a sufficient number of its features to be … what? The BAP concept appears so nebulous that it is difficult to devise a suitable descriptor for people supposedly in this category other than 'member of the BAP', which says nothing. How many of the criteria in an autism screening cluster would a person need to qualify for membership of a BAP rather than them being undeniably non-autistic, and what would the cut-off point be for actually being autistic? Including certain traits in a diagnostic cluster for autism does not mean that individuals with some, but not all, of these traits are in a 'somewhat, but not fully, autistic' category. Those who believe this have been bewitched by the hidden transition from 'human trait associated with autism' to an 'autistic trait' that implies a degree of autism,[13] whatever this may mean. Human beings can present with any combination of human traits. So human traits in a diagnostic cluster for autism may be seen in non-autistic people. This does not imply that these individuals are not autistic enough to justify a diagnosis of autism, whatever this may mean. It is simply that they have some of the human traits used to diagnose autism because at present we have no better means of diagnosing autism than by using (a cluster of) behavioural criteria (ibid.).

Confusing language

Here again, let's consider two examples. First, there is an example of the rei-fication of a piece of confusing language in autism that Bovell has discussed. She points to the crucial distinction between treating a co-morbidity and treating autism itself, writing that 'in the treatment vs acceptance debate, much of the defence of the pro-treatment group rested on their emphasis on co-morbidities that could/should be treated, and their rejection of the idea that painful co-morbidities should be 'accepted' rather than challenged' (2015, p. 275). As mentioned ear-lier, this position would be accepted by most autistic advocates. Pro-treatment groups usually call for treatment of associated health needs, not autism itself. But 'treating autism' is a far handier descriptor than, say, 'treating medical conditions associated with autism'. In other words, what began life as a headline-grabbing form of words designed to attract attention, can become something it was never intended to be, that is, a statement that autism *itself* should be treated. Of course, 'treating autism' may mean exactly that in some cases; my point is that it sends a wrong message when it does not mean what it says.

The second example is also sourced from Bovell and is an example of how crude language can oversimplify debate by concealing its underlying complexity. In relation to the impact of the problematic aspects of autism on families, such as sleep deprivation and challenging behaviour,[14] she stresses that

> Given the heterogeneity of autism, and indeed of families, there are multiple different narratives in which problems … either do not feature, or feature only at a particular point in time, and which in any case are perceived as being com-pensated for by some of the benefits that an autistic family member will bring.[15]
>
> (Ibid., p. 288)

She refers to the crude 'disabled vs non-disabled' debates relating to autism that serve to conceal the complexity resulting from such heterogeneity. In the same way that I have drawn attention to the complexities between NT and autistic per-spectives on autism, scholars should avoid crude binaries that cannot reflect the heterogeneity of attitudes to autism.

Conclusions

Overcoming 'an instance of moral blindness – when one comes to see the moral salience of something one did not see before' – requires moral perception

(Wisnewski, 2007, p. 123). Baumann (in Baumann & Donskis, 2013) refers to situations where issues become separated from concerns about their morality, which enables acts to be committed without those involved facing any moral stigma or needing to worry their consciences about them. It is my contention that the failure of most scholars working towards the cure/prevention of autism to openly discuss the ethics of their research, and the failure of society to call these scholars to account, is an example of both the separation Baumann refers to and a failure of moral perception. University ethics committees should cover the macro issue of whether or not curing/preventing autism is morally acceptable as well as the usual micro issues. Society should insist on full debates about *all* ethical issues relating to autism research.

No valid case has yet been made that the health of the social body requires the amputation of the autistic parts of the body (cf. Bauman & Donskis, 2013). Work to remove autism from the social body should not proceed in the dark space of a moral vacuum; such a fundamental issue must be brought out into the clear light of day. To ensure ethical matters in autism research are given the attention they are due, I recommend that:

1   all autism research projects should undertake an ethical impact assessment (EIA) for consideration by the university's ethics committee;
2   university ethics committees should make these impact assessments, and their deliberations on them, publicly available.

These recommendations are in line with the approach taken to ethical matters by the Human Brain Project (HBP) which 'Recognizing that its research may raise various ethical … issues has made the identification, examination, and management of those issues a top priority' (Salles et al., 2019, p. 380). The issues referred to include the values that inform, and the *ethical permissibility* of, research.

Virginia Bovell (2015, p. 86) writes that 'a crude perspective on autism, either as something that is bad and should be eliminated, or as something that is good that should be celebrated, does not do justice to the complexity of human experience'. The challenges faced by some autistic people, and their carers, must not be ignored. But we should also reflect on the fact that non-autistic individuals can pose serious challenges. In the same way that I have no qualms in saying out loud that the apparently non-autistic Donald Trump presents a clear and present danger to civilised society, I consider the campaigner Greta Thunberg[16] – who has spoken[17] of the autistic strengths that she believes have enabled her to take on a climate change activist leadership role at a young age – to be a wonderful asset to society.

Acknowledgements

My thanks to Hanna Bertilsdotter Rosqvist, Robert Chapman, and Anna Stenning for their valuable input, and to Virginia Bovell whose work was a major influence.

Notes

1 https://aeon.co/ideas/how-playing-wittgensteinian-language-games-can-set-us-free
2 Rules of social interactional game-playing are not codified like the rules of cricket or chess. They are subtle, complex and generally learned via osmosis during the formative years.
3 Barnes and McCabe (2011, p. 268) ask their readers to 'reflect on whether the world is better with or without a cure [for autism]' which suggests that the incidence of autism also concerns them.
4 One of few statements in Barnbaum's book that suggests she may not support the full eradication of autism is her reference to certain studies that 'locate – a moral sense in persons with autism' (Barnbaum, 2008, p. 111). This quotation is of particular interest to me because she appears to recognise that the ability to recognise moral questions is not a matter of neurotype.
5 It is a general goal as she reserves the right for a mother to have the final decision on whether or not to give birth.
6 The *Autonomy* journal has published the letter under Julia Leatherland's sole name.
7 Wittgenstein did not use the term 'grammar' in accordance with its dictionary definition. His definition of this term refers to the rules that govern word usage. He wrote that 'grammar … has somewhat the same relation to the language as … the rules of a game have to the game' (PG, I, 23).
8 I use the term 'neurotypicality' simply to draw a distinction between majority cognition and autism.
9 Wittgenstein intended the language game concept 'to bring into prominence the fact that the speaking of language is part of an activity' (PI 23) which gives language its meaning.
10 The bi-directional difficulty in understanding was named 'double empathy' by Damian Milton.
11 Some parents are involved in the neurodiversity movement and some autistic individuals support the search for a cure for autism. But the 'autism as tragedy' trope *is* a neurotypical concept.
12 One reason for this is that many autistic people do not disclose their autism because of the stigma still associated with autism and the risk of damaging their professional careers.
13 The DSM-5 has introduced the concept of the severity of autism. The extent of the autism-friendliness of an environment influences the apparent severity of autism presentation.
14 Sleep deprivation and challenging behaviour are not restricted to autistic children.
15 As was pointed out to me by Joanna Baker-Rogers, these benefits (which can apply in the case of children with many different labels) include the love they inspire in their family, friends, and carers.
16 I do NOT argue that autistic people are only 'acceptable' to society if they have social utility.
17 https://edition.cnn.com/videos/tv/2019/02/01/amanpour-greta-thunberg.cnn

References

Anderson, J. L. (2013). A Dash of autism. In J. L. Anderson & S. Cushing (Eds.), *The philosophy of autism* (pp. 109–142). Plymouth: Rowman & Littlefield.

Barnbaum, D. R. (2008). *The ethics of autism: Among them but not of them*. Bloomington, IN: Indiana University Press.

Barnes, R. E., & McCabe, H. (2012). Should we welcome a cure for autism? A survey of the arguments. *Medicine, Health Care and Philosophy, 15*(3), 255–269.

Bauman, Z., & Donskis, L. (2013). *Moral blindness: The loss of sensitivity in liquid modernity*. Cambridge, UK: Polity Press.

Bovell, V. (2015). *Is the prevention and/or cure of autism a morally legitimate quest?* (Unpublished doctoral dissertation). University of Oxford.

Chapman, R. (2018). *Autism, neurodiversity, and the good life: On the very possibility of autistic thriving* (Unpublished doctoral dissertation). University of Essex.

Chapman, R. (2019). Neurodiversity theory and its discontents: Autism, schizophrenia, and the social model of disability. In R. Bluhm (Ed.), *The Bloomsbury companion to philosophy of psychiatry* (pp. 371–390). London: Bloomsbury Academic.

Chown, N. (2014). More on the ontological status of autism and double empathy. *Disability & Society, 29*(10), 1672–1676.

Chown, N. (2019). Are the 'autistic traits' and 'broader autism phenotype' concepts real or mythical? *Autism Policy and Practice, 2*(1), 46–63.

Chown, N., & Leatherland, J. (2018). An open letter to Professor David Mandell, Editor-in-Chief, 'Autism' in response to the editorial 'A new era in autism'. *Autonomy, 1*(5). Available at: www.larry-arnold.net/Autonomy/index.php/autonomy/article/view/CO1/html

Drescher, J. (2015). Out of DSM: depathologizing homosexuality. *Behavioral Sciences, 5*(4), 565–575.

Hughes, J. M. F. (2015). *Changing conversations around autism: A critical, action implicative discourse analysis of US neurodiversity advocacy online* (Unpublished doctoral dissertation). University of Colorado.

Kapp, S. K. (Ed.) (2020). *Autistic community and the neurodiversity movement: Stories from the frontline*. London: Palgrave Macmillan.

Milton, D. E. (2012). On the ontological status of autism: the 'double empathy problem'. *Disability & Society, 27*(6), 883–887.

Milton, D., & Bracher, M. (2013). Autistics speak but are they heard? *Medical Sociology Online, 7*(2), 61–69.

Murray, D., Lesser, M., & Lawson, W. (2005). Attention, monotropism and the diagnostic criteria for autism. *Autism, 9*(2), 139–156.

Nadesan, M. (2013). Autism and genetics profit, risk, and bare life. In J. Davidson & M. Orsini (Eds.), *Worlds of autism: Across the spectrum of neurological difference* (pp. 117–142). Minneapolis, MN: University of Minnesota Press.

Pellicano, E., Dinsmore, A., & Charman, T. (2014). What should autism research focus upon? Community views and priorities from the United Kingdom. *Autism, 18*(7), 756–770.

Pellicano, E., & Stears, M. (2011). Bridging autism, science and society: moving toward an ethically informed approach to autism research. *Autism Research, 4*(4), 271–282.

Pripas-Kapit, S. (2020). Historicizing Jim Sinclair's 'Don't mourn for us': A cultural and intellectual history of neurodiversity's first manifesto. In S. K. Kapp (Ed.), *Autistic community and the neurodiversity movement*. London: Palgrave Macmillan.

Salles, A., Bjaalie, J.G., Evers, K., Farisco, M., Fothergill, B. T., Guerrero, M., … & Walter, H. (2019). The human brain project: responsible brain research for the benefit of society. *Neuron, 101*(3), 380–384.

Sinclair, J. (1993). Don't Mourn for Us. *Our Voice, 1*(3). URL: (Retrieved December 2019) www.autreat.com/dont_mourn.html

Szasz, T. (2010). *The myth of mental illness: Foundation of a theory of personal conduct*. New York: Harper Perennial.

Walsh, P. (2010). Asperger syndrome and the supposed obligation not to bring disabled lives into the world. *Journal of Medical Ethics*, *36*(9), 521–524.

Wisnewski, J. J. (2007). *Wittgenstein and ethical inquiry: A defense of ethics as clarification*. London: Continuum.

Wittgenstein, L. (1984). *Culture and value*. Chicago, IL: University of Chicago Press.

Wittgenstein, L. (2005). *Philosophical Grammar*. Berkeley: University of California Press.

Wittgenstein, L. (2009). *Philosophical Investigations*. Hoboken, NJ: John Wiley & Sons.

Chapter 3

Is there an ethical case for the prevention and/or cure of autism?

Virginia Bovell

Research to find causes and treatments for autism reflects a powerful belief that autism entails difficulties that are best avoided or ameliorated. The response from the neurodiversity movement has been to question the pathologising of autism, and to challenge the medicalised culture in which diagnosis and treatment are governed by notions of 'normal' human functioning. It is argued that the difficulties experienced by autistic people come less from within autistic individuals than from the wider social and attitudinal environment around them and their interaction with it – for example through misunderstanding, neglect, stigmatising, and societal pressures to be 'normal' (Lawson, 2008), and/or failure to recognise that autistics represent a legitimate minority group with a distinct identity (Chown, 2013).

Against such a view, some parents and autistic people continue to advocate for prevention and cure (Mitchell, 2019). They are concerned that autism is misrepresented by autistic people who self-advocate in sophisticated and articulate ways, in contrast to others whose impairments are profound and debilitating and who, it is argued, are at risk of being ignored if the self-advocates become the sole depiction of autism in the priorities of science and society.[1]

This dichotomy, which has been labelled the 'treatment vs acceptance debate', features in both lay and academic publications (Jaarsma & Welin, 2013; Laurance, 2007). In this chapter I seek to go beyond theoretical positions, by offering examples of what prevention and/or cure actually entail in practice. In so doing, I aim to demonstrate there are serious limitations to arguments promoting prevention and cure, even if one approaches the issues from within a traditional medical ethics perspective. Given this, I suggest areas of common ground that might assist in moving forward from what has until now been a binary and polarised debate.

Preliminary questions

Advocates of prevention and/or cure need to be able to address several important questions:

1   In hoping that autism might one day be cured or prevented, what do they actually mean? There are significant challenges in pinning down what autism

is (Verhoeff, 2012). Increasingly, scientists are referring to autisms, plural (Waterhouse & Gillberg, 2014) while others ask whether there is any underlying biological essence to the condition(s) at all (Timimi, Gardner, & McCabe, 2011). This alone should make us question how we can coherently call for the cure or prevention of something that remains so elusive.

2   Whose perspective on autism is most relevant in determining what are ethical intervention targets: those who emphasise impairments and problems linked with autism, or those who challenge the idea of autism as pathology and who emphasise its strengths?

3   Does the heterogeneity of autism permit a differential approach towards prevention and cure, depending on what 'type' of autism is being addressed? For example, Jaarsma and Welin, 2013, argue in favour of social accommodation for those with 'mild autism' but maintain that it is right to seek to prevent 'severe autism' (their words).[2]

4   Alternatively, are there integral attributes among all autistics that are relevant in evaluating the morality of seeking prevention and/or cure?[3]

Answering these broad questions requires an exploration into potential and real-life scenarios in which ethical positions are applied.

Scenarios for ethical debate

In exploring interventions into autism, there are two dimensions of ethical significance:

1   The dimension of *timing* – from measures that might be undertaken before a child is conceived through to measures that might be introduced once an autistic individual is living in the world;

2   The *target of intervention* – from biological measures influencing the environment in which conception takes place, through measures that attempt to alter the physiology or psychology of an autistic individual, through to measures that seek to change the social and material environment in which autistic people live.

Combining these dimensions delivers six categories for exploration: three that relate to interventions pre-birth, and three that apply once someone has been born. Each will be explored in turn.

A: Antenatal intervention

AI – Pre-conceptual intervention

Ethical questions that fall under this heading include: whether or not the intervention is imposed upon, or chosen by, prospective parents; what consideration should be given to the welfare of any resulting child; and the wider impact of the decisions made.

A1.1 – Removal of environmental cause

Suppose that a particular pesticide Brand X has been shown to cause autism. Preventing autism would be through the withdrawal of Brand X, either voluntarily by the manufacturer, or through government enforcement. Whereas most pre-conceptual measures are private and individually chosen, this scenario is externally imposed on prospective parents and future children. It affects the whole population within the area of pesticide use, thereby overriding any private or personal opinions about autism that prospective parents may have. Whether or not this is a problem, and why this depends on how autism is viewed, is illustrated if we consider two contrasting analogies.

1 If autism is viewed as a profoundly damaging condition, with the pesticide analogous to the thalidomide drug, then preventing autism via removing the environmental trigger seems to be morally acceptable; removing private choice from the prospective parents is of little concern.

2 If, on the other hand, Brand X leads to specific personality dispositions that carry the potential for both harm and benefit, it will be harder to achieve consensus on the right response. An example could be a disposition that predisposes individuals to exceptional courage bordering on the foolhardy. This character trait might contribute both to increased rates of hospitalisation and death for those affected, and yet might also generate admiration and appreciation from society, with such people contributing positively to valuable services such as fire-fighting. Removing Brand X might then lead to debates about the positive and negative features of such personalities, the rights of parents to choose, and the wider impact on society of the decision.

Clearly, these are not equivalents with autism. But they do illustrate a number of ethical considerations – about the significance of parental autonomy, the impact on a potential child, and wider social consequences – all of which are sensitive to ways of viewing autism – as a harmed state, or as a type of person.

A1.2 – Rejection of potential 'carrier' eggs

In 2009, a potential egg donor was rejected by four fertility clinics in England, because one of her children had Asperger's (Keeler, 2009). The decision pre-empted any opportunity for potential recipients to exercise autonomy, and against the express wishes of the potential donor. Keeler pointed out the wider impact of the decision: 'There is an acute shortage of altruistic egg donors; in rejecting me the message is that it is better to be childless than to have a child with AS' (2009).

Whether intended or not, this message conveys a damning view about the value of autistic lives, and feeds into the objections that disability rights advocates raise about the damage caused by discrimination in antenatal practices. This is an example of an important issue within medical ethics – the so-called 'expressivist' perspective. It holds that we have to take seriously the way localised decisions can have adverse broader consequences. If one person's decision impacts on the welfare of others – for example through reinforcing stigmatisation and discrimination against living disabled people – it becomes more than an isolated act and has broader moral implications.[4] Once again, the perspective on autism is key here. Revisiting the contrasting analogies of thalidomide and the distinct personality disposition, Keeler's stance favoured the latter: 'When my daughter struggles, she does so considerably, however when she flies, she soars. I wonder if it is either possible or desirable to breed out these extreme states from our species' (2009).

A2 – Antenatal prevention involving screening and selection

This scenario is distinct from A-1 because it involves an already-existing potential child. Selection can take place by avoiding implantation of an embryo, or by abortion.

A2.1 – Selective implantation

This scenario anticipates the use of pre-implantation genetic diagnosis (PGD) to identify and discard embryos possessing genes or clusters of genes that raise the chance of autism.[5] There are three areas of ethical relevance. First, a clash of principles may apply if parental autonomy leads to selecting against autism, even though the act of selection sends out a broader, stigmatising message to autistic people that they are not welcome in the world. Second, the analogy challenge recurs, with respect to whether or not autism fits within the 'serious inherited disease' category for which PGD is currently practised (HFEA, 2019). Third, some might justify selecting against autism for the sake of family balancing (for example where the family already have two autistic children – analogous to couples seeking a girl, having had two boys).

A2.2 – Abortion of at-risk foetuses

There is a vast literature within bioethics around abortion. Aside from the pro-life vs right-to-choose debate, the disability rights perspective criticises the fact that discriminatory exceptions are often made around abortion in the case of potential disability. Here, as before, the issue of messaging plays an important part in the ethical discussions. It links to the fundamental issue of what kinds of human life are considered to be valid and of equal status to others, and who has the power or authority to come to such judgements.

A3 – Antenatal prevention influencing foetal trajectory

A-3 offers an alternative method of prevention for those who shun termination, but it does require an intervention affecting an existing foetus. I offer two contrasting examples, based on candidate theories about the causes of autism. The issue of intentionality is of key ethical relevance.

A3.1 – Routine health promotion measures

Under this heading sit the types of intervention that a pregnant woman would be advised to take as 'common sense' based on existing knowledge of sound antenatal practice and general health promotion. For example, some research suggests that taking folic acid in pregnancy will reduce the chances of having a child with autism (Schmidt et al., 2013). Here, preventing autism may arise merely as a by-product of a more general goal to promote maternal and infant health.

Most, but not all, commentators believe that the issue of motive and intentionality are morally relevant in such a case, and that just happening to prevent autism is ethically unproblematic, in a way that deliberate attempts are not.

A3.2 – Targeted medical intervention

Pharmacological treatments while the child is in utero have been postulated in order to prevent autism (Zimmerman & Connors, 2014). Aside from the risks and potential adverse side effects, this type of intervention impacts on something of significance to many autistic advocates: that of *identity*. For those who hold autism to be an integral part of someone's essence, and/or confers a group identity, then the intervention is analogous to tampering with a potential child's intrinsic nature, or trying to alter someone's sexual orientation before they are born.

Even this brief exploration of scenarios is sufficient to show that prevention is, at the very least, questionable in moral terms. It is not the self-evidently good thing that is implied in much of the scientific discourse.

P: Post-birth intervention

Here, the underlying ethical question, as polarised in the treatment vs acceptance debate, is whether – in the spirit of seeking to help autistic people – interventions should target autism itself, or alternatively the wider environment, or a mixture of both. For those who argue that autism requires more than environmental change, what is the most helpful intervention? Is it one that aims to address autism *in toto*, or is it one that seeks to provide targeted support for discrete aspects of difficulty? If the latter, are such difficulties framed in terms of deviations from neurotypical behaviour, or in terms of distress experienced and prioritised by the individuals themselves? The distinctions between these propositions are captured in the three categories of postnatal intervention that I will now outline in turn.

P1 – Postnatal intervention altering an individual's trajectory away from autism, including the expressed possibility of prevention, cure, and recovery

Prevention

Dawson (2008) has asserted that very early pharmacological and behavioural interventions could prevent autism. 'Prevention will entail detecting infants at risk before the full syndrome is present and implementing treatments designed to alter the course of early behavioral and brain development' (p. 775).

She also talks about 'restoring' regulation (2008, p. 775), which implies a narrative of autism in which normality was the original design and purpose, and that autism intervened as a deviation from normal development. This is a common view, yet others question this idea that autism somehow takes over a normal person; they regard autism as integral from the start – that is, autism is 'normal' for that individual from the off.

Cure

Descriptions of autism as 'incurable', along with sensationalist headlines about miracle cures, present autism as something for which cure is self-evidently a good goal. Likewise, the term *recovery* conveys an idea of autism as an undesirable and unhealthy state, from which – with psychosocial and education interventions – a person can move on and leave their autism behind.

P1.2

Aside from questions about what interventions are aiming to achieve, there are several ethical questions about practice.

- Who ought to make the choice about whether or not to seek cure/prevention?
- When and why is it made?
- What is involved in the intervention?
- How likely is it to be effective and how is effectiveness measured?
- Does it involve discomfort?
- Does it entail adverse side-effects, in the immediate through to the long term?
- What are the autonomy and consent issues at stake (given the age and/or intellectual capacity of the recipients of the intervention)?
- What moral issues arise for an intervention that may impact on the sense of identity of the developing person?

- At what stage of development is it right to instigate and/or abandon these types of intervention?
- Are there resource implications, for example if priority is given to focused interventions on the young, thereby reducing the support available for autistic individuals who are older?
- If a 'cost of care' argument is made in favour of diverting children away from autism, will families be pressurised into adopting these interventions even if they believe them to be wrong for their child?
- How is the choice framed to the individual affected?

This list is relevant in most kinds of interventions, but is particularly important if something as dramatic as cure is being considered, and if the person does not have capacity to consent.

P2 – Postnatal intervention focused on amelioration of an individual's difficulties

The goals of P-2 are more modest than P-1, but many of the above questions remain relevant. P-2 interventions rest on the idea that the most helpful thing to do is to equip autistic people with strategies to overcome at least some of their difficulties, taking the impossibility of curing autism (removing it *in toto*) as a given. While accepting that autism is a lifelong condition, regardless of intervention, P-2 interventions may be applied at different stages throughout a person's life, depending on presenting issues and challenges.

P3 – Postnatal intervention focused on external change – social, cultural, and policy responses to autism

P-3 is aligned with the social model of disability and the neurodiversity movement, and also takes into account the views of those parents for whom the key struggles of having an autistic child relate to accessing support: 'It's not X's autism that's the problem, it's the battle for services' (anon, 2001, pers. comm.).

Interventions that fall within P-3 relate to this wider dimension, covering measures that aim to alter the social, institutional, and political context in order to improve autistic people's chances of full engagement in a diverse society, without targeting the behaviour or biological functioning of autistic people themselves.

Examples include government strategies and generous levels of funding in education and social care, through to specific laws such as the UK Autism Act (Great Britain, 2009). P-3 interventions emphasise the need for behaviour change on the rest of the population – for example, anti-bullying campaigns and the legal requirement in the UK for institutions to make 'reasonable adjustments' (Great Britain, 2010).

P3.1 – Ethical goals vs real-world choices

While P3 interventions might be broadly supported in principle, working towards societal change is a complex and long-term project. Meanwhile, people are faced with local and immediate choices about how best to live as, or be supportive to, autistic individuals. For this reason, there may be a mismatch between what people would like to see, and what they believe is achievable:

- There may be parents, practitioners and autistic people who long for P-3 but who feel helpless to change society, and therefore focus on P-2.

Conversely:

- There may be P-2 professionals, parents, and academics who would see nothing wrong in the goals aligned with P-1, but believe this is not achievable, or not within their sphere of practice. They would like to see P-1 but, believing this is not possible, settle for P-2.

And moreover:

- There may be people who talk as if they want P-1, but who are actually closer to P-2 and P-3.

I believe that it is this last category of people who fall foul of the double bind of scientific uncertainty about what – if anything – autism really is, and semantic confusion. Regarding semantic confusion, I consider there to be an important distinction that is often overlooked between the core features of autism as defined in the diagnostic criteria – DSM-5 (American Psychiatric Association, 2013), and the issues that some autistic people – or their families – find particularly distressing. In this last category, people may call for cure when actually they are talking about co-occurring conditions: the DSM 'specifiers' that only apply to some autistic people. Additional semantic ambiguities – such as using one term (e.g. 'recovery', 'normalisation', 'treatment') to mean more than one thing – may exacerbate polarisation. Greater precision in concepts and terminology is likely to narrow down and clarify the remaining areas of dispute.

P3.2 – Revisiting the 'treatment vs acceptance' polarity

The treatment vs acceptance debate is positioned as a P-1 vs P-3 debate, and one in which there is no common ground. Yet this ignores P-2 and also the potential for overlap and interdependence between the categories. In an attempt to clarify both the distinctions and the interdependencies, I now set out three propositions.

P3.2.I – 'TREATMENT VS ACCEPTANCE' IS A FALSE DICHOTOMY

If one believes that interventions focused on the individual are necessary or ben-eficial, then it follows that they should be available, accessible, and delivered to a high standard of professional practice. This is likely to require investment and political will at a system-wide, P-3, level. As such, one would expect P-3 interventions to gain the greatest amount of support from across the stakeholder community. Yet if a change in the wider institutional environment does trigger large-scale investment, the downstream impact may well be to increase or improve individually-targeted (P1 or P2) services.

P3.2.2 – P-2 INTERVENTIONS COULD BENEFIT FROM GREATER DIALOGUE AMONGST STAKEHOLDERS AS TO THEIR NATURE, PURPOSE, AND EMPHASIS

Most autistic advocates are not opposed to interventions to assist with particular areas of difficulty, where these are person-centred and sensitive to context. What autistic advocates have criticised to date is the *dominance* of the individual-based, behaviour-change approach, where the onus for adaptation has been entirely one way – that is, all the pressures for the change apply to autistic people, with no reciprocal adaptation from NTs. In addition, they question the focus on core autism symptoms, rather than other targets that are a priority for autistic people and those close to them.

P3.2.3 – AMONG POSTNATAL INTERVENTIONS, P-I INTERVENTIONS POSE THE GREATEST ETHICAL PROBLEM

Here the polarisation appears to reflect a conceptually irreconcilable difference. P-1 interventions seek to remove autism, 'and therefore autistic people' (Dawson, 2004), *in toto*. Whether or not P-1 advocates *intend* to make such a sweeping point does not detract from the implications of the message. It may be that when faced with the implications of P-1, they opt to move away and say that they favour effective and respectful interventions within P-2 and P-3 – measures that can ena-ble autistic people to flourish and to fulfil their potential according to their varied needs and abilities within a more supportive environment.

Alternatively, P-1 advocates might argue that they are not calling for a future world in which there are no more autistic individuals, but only for a future world in which people can *choose*, or not, whether to opt for treatment/prevention measures. The problem with this second position is that in virtu-ally all the scenarios referred to – pre-birth or in infancy – the potentially autistic person does not make the choice, and parents may not be fully aware of the implications. 'Prospective parents are dealing in the abstract with some-thing that could become tangible, and that's never an informed way to make a choice' (Solomon 2013, p. 29).

Discussion

Opinions on how best to serve the interests of autistic people therefore embrace views both on what autism is and also what kind of world we live in. It is important to recognise the impact of – and interplay between – these two belief systems.

Autism as departure from an idealised 'norm'

Medical ethicists have referred to the idea of 'normal species functioning' when seeking a shared understanding of a minimum quality of life to which all humanity should aspire (Boorse, 1975). Yet there is no prima facie argument for assuming that departures from the norm are necessarily incompatible with a good outcome. Ari Ne'eman (cited in Solomon, 2013, p. 275) has made the point that in strict statistical terms, normal merely means 'average' and further that the midpoint on the bell curve signifies nothing more than 'mediocrity'. Some departures from the norm confer advantage, while normal species standards may overlook the interests of all kinds of minorities (Silvers, 1998).

Autism as identity

While some refer to autism chiefly in terms of enhanced or diminished abilities in different areas, others conceptualise autism chiefly in terms of identity. If autistic people are right to claim to be a distinct minority group, then this requires a different ethical position from one based on a view that autistic people are individuals who have all fallen foul of a specific disorder. In repudiating the latter idea, Jim Sinclair's foundational position about autism as identity asserts: 'Autism is a way of being. It is *pervasive*; it colors every experience, every sensation, perception, thought, emotion and encounter, every aspect of existence. It is not possible to separate the autism from the person' (Sinclair, 1993).

Many other autistic voices echo this perspective, along with Sinclair's assertion that there is a distinct autistic community (Sinclair, 2005), and that autistic people constitute a minority group with shared interests (Chapman, 2020; Chown, 2013).

Yet others disagree, expressing the view that autism is a separate thing from someone's whole identity, and appear not to worry that cure would require a break in their narrative continuity (Mitchell, 2019).

Both views can co-exist as personal beliefs, but not as wider policy approaches. Thus Scully (2008) has argued that even if large numbers were to welcome prevention or cure for themselves or their loved-ones, this does not legitimate it as a generalised goal. It is not known how much inherent disadvantage would be left once the societal disadvantages surrounding the condition have been removed. In other words, until we have exhausted the possibilities of P-3 and P-2 interventions, it is premature to consider the justification of antenatal or of P-1 interventions.

Perspectives on harm and suffering with regard to autism

Medical ethicists draw heavily on a welfare calculus, weighing up harm and benefit, with an emphasis on avoiding or alleviating suffering. It is important therefore to see if the equation of autism and suffering stands up to scrutiny.

The impact on individuals

Long-term follow-up studies suggest that autistic people have rather poor outcomes, if these are measured by standard socio-economic and lifestyle indicators (Howlin, Moss, Savage, & Rutter, 2013). These data might be regarded as evidence of suffering, but they need to be viewed with care. Autistic writers tend to offer contrasting emphases when speaking about their own lives. Many of the most distressed accounts come from the experience of additional challenges such as sensory sensitivity and anxiety, not to the core features of autism. Their arguments suggest that they do not in any case suffer more than anyone else and/or that any suffering is worth it because of compensating enhanced pleasures and perceptions, and/or that their lives are worthwhile regardless of attempts to measure suffering, and/or that the key causes of suffering come from the wider environment and not from autism itself. While many indicate that their autism entails difficulties, many or perhaps most would argue that this is not such a problem as to justify taking away their autism, nor to have prevented their birth (see, for example, Robison, 2007 and 2013).

So we need to question the assumption that to prevent autism is good because people with autism suffer – even if it is true that they (or some) do suffer. Such suffering needs to be counterbalanced against the potential harm and suffering that might arise from attempts to cure and prevent autism. Also, exploration is needed with regard to how quality of life might be conceptualised and measured in autism-sensitive ways (Robeyns, 2018; Rodogno, Krause-Jensen, & Ashcroft, 2016).

Medical co-morbidities

The chances of both epilepsy and gastro-intestinal illness are raised in autism, both of which can debilitate at best and lower life-expectancy at worst. These are medical problems *within* the person, and this marks them out as different from difficulties that exist primarily due to and/or only within the wider inter-personal and societal context.

Therefore – in our current state of knowledge – knowingly to enable the birth of an autistic person is knowingly to raise the chances of this new person experiencing painful and occasionally life-threatening co-occurring conditions. However, this does not provide a general justification for preventing all autistic births – any more than one would call for a reduction in ovarian cancer by calling for fewer girls to be born. Because some groups are more prone to certain conditions does not justify eliminating the group. If we view autism's co-occurring conditions in this way, then the moral and practical challenge is to address the conditions – which is not at all the same thing as addressing autism itself.

Non-medical 'co-morbidities'

It is often implied that autistic people who are intellectually able and verbal are capable of having a good life and have more to contribute to the world, while the 'low functioning' live sub-optimal lives (however this is defined). Yet this is questionable both empirically and morally.

Autistic people can be extremely cognitively able and yet have major difficulties in other domains. And where suffering does occur for the person with learning disability (LD), this is often contingent upon social conditions such as stigma, rather than an inherent disadvantage. People with LD are often portrayed as somehow less than fully human (Carlson 2010; Kittay & Carlson, 2010), while assumptions about suffering may merely reflect a wider societal aversion to certain types of dependency. Thus Thomas (2007 p. 88) contrasts a disabled person's reliance on help to dress, wash, and cook, with a globe-trotting businessman's reliance on pilots, hotel staff, and his wife/domestic staff. The key difference is one of perceived autonomy and power, which suggests that (in) dependence itself is less significant than the vulnerability that comes with certain types of dependence.

If we look at vulnerability – to bullying, exploitation, hate crimes, – it is not clear which subgroup within autism is most affected. Where adequate love and support are provided, then for the LD individual, the disadvantages/stigma of being 'dependent' may be far outweighed by the security and protection that being supported can bring. In the wider narratives, very little attention is paid to the positive aspects of life among the LD autistic population: while the reality may be devastating for some, it is inaccurate and misleading to universalise the bad experiences.

The impact of autism on families

Disability rights advocates recognise that there are specific challenges that some parents of disabled children face, such as lack of sleep, violent behaviour, the pain of having to give up direct care for the child, or the opposite challenge of unremitting care with no external support (Solomon, 2013; Terminiello, 2019).

Yet, given the heterogeneity of autism, and, indeed, of families, there are multiple narratives in which problems feature only at a particular point in time, and which, with appropriate and personalised support, do not signify a lifetime of blight – rather, the reverse. With greater practical support, and in an atmosphere that is less stigmatising and adversarial, many of the problems affecting families can be allayed. And this comes back to the role of P-2 and P-3 interventions, such as a social climate that celebrates diversity, greater willingness to fund early support, education, high quality social care, supported employment, and so on – all reinforced by investment in research and practice to improve the quality of life of future generations.

The impact of cure/prevention on the wider population

In defence of autistic people's positive contribution to the world, much is often made of the talents possessed by some individuals; increased recognition is given to autism not just in terms of deficits, but also in terms of strengths.

While it is important to acknowledge autistic strengths and to challenge the traditional emphasis on deficits, this would be a dangerous sole basis on which to justify autistic people's existence. Assessing the value of a life on a person's ability to earn, to perform or achieve in exceptional ways is questionable at an individual level, and is actively pernicious when applied to groups. Further, there are several intangibles and externalities that lead individuals, families, and communities to be enriched in ways that cannot be calculated in terms of monetary or wider social recognition.

Similarly, we should be very careful when interpreting statements about the overall economic costs of autism, mindful of the disability movement's critique of the bias implicit in singling out the disabled when making public-expenditure arguments. The choice of what are considered to be necessary and normal areas of public expenditure, as opposed to avoidable and burdensome areas, is a political one that reflects underlying power structures and priorities about who are legitimate participants in our society – and who are not. For example, we do not call for prevention/cure of people who are physically active, despite expenditure on sports facilities and medical treatment of injuries.

Moreover, cost data alone do not take into account the macro-economic significance of social, education, and health-care spending more broadly. The bulk of autism's so-called costs are largely allocated to employ people who in turn pay taxes and/or spend their earnings and stimulate aggregate monetary demand. The funding allocated to autistic people operates within the circular flow of income, rather than being an extraneous black hole that merely swallows up money and puts nothing back.

Conclusion

Behind calls to prevent or cure autism there is a complex web of issues, all of which carry significant moral implications. The extent of this complexity alone should make people pause before offering simple mission statements. The first complexity relates to autism itself. Scientists are moving away from seeing autism as a unitary, reducible condition, or as one that allows simple predictions about what kind of life is implicated for an autistic person or for those around them. With respect to post-birth interventions, I have suggested that at least some of the challenges that autistic people and their families face are due to wider social, cultural and institutional conditions, rather than intrinsic to autism itself. This shifts the target of ethical intervention towards wider measures – namely P-3 interventions. However, while P-3 interventions are the least morally problematic, they are nonetheless compatible with favouring individually targeted interventions, as long as these are P-2 rather than P-1. In other words, I believe the purpose of such assistance is only morally acceptable when conceived as a method for achieving wellbeing for the individual (and those close to them), and should not be conflated with the goal of taking away their autism. In the case of pre-birth interventions, I am not ready to override the right of the mother to be the ultimate decider about her private reproductive choices, notwithstanding the need to challenge at a broader level the normalising assumptions and messages that pervade antenatal practice (Shakespeare, 2005). This is in line with feminist and disability rights advocates' support of maternal autonomy. My overall conclusion is that reference to prevention and/or cure as a desirable *general* goal is neither clinically/scientifically coherent nor morally legitimate. To talk in approving terms about prevention and cure implies that a world where there are no more autistic people would be a better world. Aside from how offensive this is to some, it is wildly simplistic. To persist in justifying prevention and cure as broad goals means to persist in believing not only that such a future world would be a better place but also that it is acceptable to adopt this terminology despite the offence and anxiety it causes to at least some autistic people, and to many who love them. If, on the other hand, what is hoped for is *choice*, and only in *limited* circumstances, then this should be made crystal clear. That does not end ethical debate, but it does, I hope, move us away from polarised positions around 'treatment vs acceptance', and clear the ground for more nuanced, relevant, and context-sensitive conversations.

Acknowledgements

I am very grateful to many friends and colleagues whose views have influenced me over the years, and in particular to Dr Larry Arnold and Dr Dinah Murray for extensive discussions on these issues.

Notes

1 A linked concern is that scientific enquiry is increasingly focused on the more 'high functioning' (Russell et al., 2019).
2 Use of the terms 'severe' and 'mild', and the notion of a binary distinction between them, are highly debatable, and are criticised by many commentators.
3 For example, capacity for empathy has been discussed with this in mind (Krahn & Fenton, 2009).
4 For several essays on pre-natal testing and selection, and the expressivist critique, see Parens and Asch (2000).

5 Cruder ways of preventing autism via PGD already exist. In Western Australia it has been considered acceptable to reject all male embryos because autism occurs four times more often in boys than girls (Brice, 2013).

References

American Psychiatric Association (2013). *Diagnostic and statistical manual of mental disorders* (5th ed.). Washington, DC: American Psychiatric Publishing.

Anon (2001). Mother of an autistic child. Conversation with the author. Personal communication.

Boorse, C. (1975). On the distinction between disease and illness. *Philosophy and Public Affairs*, *5*(1), 49–68.

Brice, P. (2013). *Embryo selection to prevent autism approved in Australia*. 22 October. URL: (Retrieved July 2019) www.phgfoundation.org/news/14834/

Carlson, L. (2010). *The faces of intellectual disability: Philosophical reflections*. Bloomington, IN: Indiana University Press.

Chapman, R. (2020). The reality of autism: on the metaphysics of disorder and diversity (*Philosophical Psychology*, forthcoming).

Chown, N. P. (2013). The mismeasure of autism: a challenge to orthodox autism theory. *Autonomy, the Critical Journal of Interdisciplinary Autism Studies*, *1*(2).

Dawson, M. (2004). The misbehaviour of behaviourists: ethical challenges to the autism-ABA industry. URL: (Retrieved July 2019) www.sentex.net/~nexus23/naa_aba.html

Dawson, G. (2008). Early behavioral intervention, brain plasticity, and the prevention of autism spectrum disorder. *Development and Psychopathology*, *20*(3), 775–803.

Great Britain. (2009). *Autism Act 2009*. London: The Stationery Office.

Great Britain. (2010). *Equality Act 2010*. London: The Stationery Office.

HFEA (2019). *Pre-implantation genetic diagnosis (PGD)*. URL: (Retrieved July 2019) www.hfea.gov.uk/treatments/embryo-testing-and-treatments-for-disease/pre-implantation-genetic-diagnosis-pgd/

Howlin, P., Moss, P., Savage, S., & Rutter, M. (2013). Social outcomes in mid- to later adulthood among individuals diagnosed with autism and average nonverbal IQ as children. *Journal of the American Academy of Child and Adolescent Psychiatry*, *52*(6), 572–581.

Jaarsma, P., & Welin, S. (2013). Human capabilities, mild autism, deafness and the morality of embryo selection. *Medicine, Health Care and Philosophy*, *16*, 817–824.

Keeler, H. (2009). Unnatural selection. *BioNews*, *538*, 14 December.

Kittay, E. F., & Carlson, L. (Eds.) (2010). *Cognitive disability and its challenge to moral philosophy*. Chichester: John Wiley.

Krahn, T. & Fenton, A. (2009). Autism, empathy and questions of moral agency. *Journal for the Theory of Social Behaviour*, *39*(2), 145–166.

Laurance, J. (2007). Autism: What are the ethics of treating disability? A row about the correct response to the condition – acceptance or treatment – is dividing campaigners and carers. *Independent*, 16 November.

Lawson, W. (2008). *Concepts of normality: The autistic and typical spectrum*. London: Jessica Kingsley.

Mitchell, J. (2019). The dangers of 'neurodiversity': why do people want to stop a cure for autism being found? *The Spectator*, 19 January.

Parens, E., & Asch, A. (Eds.) (2000). *Prenatal testing and disability rights*. Washington, DC: Georgetown University Press.

Robeyns, I. (2018). Conceptualising well-being for autistic persons. *Journal of Medical Ethics*, *42*, 383–390.

Robison, J. E. (2007). *Look me in the eye: My life with Asperger's*. London: Ebury Press.

Robison, J. E. (2013). I resign my roles at Autism Speaks. *John Elder Robison blog*. Posted 13 November 13. URL: (Retrieved July 2019) http://jerobison.blogspot.coluk/2013/11/i-resign-my-roles-autism-speaks.html

Rodogno, R., Krause-Jensen, K., & Ashcroft, R. E. (2016). 'Autism and the good life': a new approach to the study of well-being. *Journal of Medical Ethics*, *42*, 401–408.

Russell, G., Mandy, W., Elliott, D., White, R., Pittwood, T., & Ford, T. (2019). Selection bias on intellectual ability in autism research: a cross-sectional review and meta-analysis. *Molecular autism*, *10*(1), 9.

Schmidt, R., Tancredi, D. J., Ozonoff, S., Hansen, R. L., Hartiala, J., Allayee, H., Schmidt, L. C., Tassone, F., & Hertz-Picciotto, I. (2013). Maternal periconceptional folic acid intake and risk of autism spectrum disorders and developmental delay in the CHARGE (CHildhood Autism Risks from Genetics and Environment) case-control study. *American Journal of Clinical Nutrition*, *96*(1), 80–89.

Scully, J. L. (2008). *Disability bioethics: Moral bodies, moral difference*. Lanham, MD: Rowman & Littlefield.

Shakespeare, T. (2005). The social context of individual choice. In D. Wasserman, J. Bickenbach & R. Wachbroit (Eds.), *Quality of life and human difference: Genetic Testing, health care, and disability* (pp. 217–236). Cambridge: Cambridge University Press.

Silvers, A. (1998). A fatal attraction to normalizing: Treating disabilities as deviations from 'species-typical' functioning. In E. Parens (Ed.) *Enhancing human capacities: Conceptual complexities and ethical implications* (pp. 95–123). Washington, DC: Georgetown University Press.

Sinclair, J. (1993). Don't mourn for us. *Our Voice*, *1*(3). URL: (Retrieved December 2019) www.autreat.com/dont_mourn.html

Sinclair, J. (2005). *Autism Network International: The development of a community and its culture*. URL: (Retrieved July 2019) www.autreat.com/History_of_ANI.html

Solomon, A. (2013). *Far from the tree: a dozen kinds of love*. London: Chatto & Windus.

Terminiello, J. (2019). Opinion: The real autistics are being left behind. *Burlington County Times*. 22 June. URL: (Retrieved July 2019) www.burlingtoncountytimes.com/opinion/20190622/guest-opinion-real-autistics-are-being-left-behind

Timimi, S., Gardner, N., & McCabe, B. (2011). *The myth of autism: Medicalising men and boys' social and emotional competence*. Basingstoke: Palgrave Macmillan.

Thomas, C. (2007). *Sociologies of disability and illness: Contested ideas in disability studies and medical sociology*. Basingstoke: Palgrave Macmillan.

Verhoeff, B. (2012). What is this thing called autism? A critical analysis of the tenacious search for autism's essence. *BioSocieties*, *7*, 410–432.

Waterhouse, L., & Gillberg, C. (2014). Why autism must be taken apart. *Journal of Autism and Developmental Disorders*, *44*(7), 1788–1792.

Zimmerman, A. W., & Connors, S. L. (2014). Could autism be treated prenatally? *Science*, *343*(6171), 620–621.

Part II

Neurodivergent wellbeing

Chapter 4

Neurodiversity, disability, wellbeing

Robert Chapman

Introduction

The still-dominant 'medical model' of disability can be understood on at least two levels. First, as a widespread ideological response to disability, pervading attitudes, policy, social structures, and representations of disability. For those who analyse it at the ideological level, the medical model has sometimes been described as the 'personal tragedy model', in so far as proponents often frame disability as 'objectively bad, and thus something to be pitied, a personal tragedy for both the individual and her family' (Carlson 2010, p. 5). On a different level it is more a specific theoretical framework for understanding disability as medical pathology. In the latter, more restricted sense, the medical model is simply that which holds disability to refer to 'conditions that are abnormal and negatively deviate from normal human species' functioning and which are therefore harmful and which we should generally try to prevent or correct' (Savulescu & Kahane, 2011, p. 45).

My focus here is cognitive disability,[1] and the notion that it is 'objectively bad'. In both theoretical and ideological representations, cognitive disability is often associated with a purported inherent disposition towards either low levels of wellbeing or limited capacity to flourish. In the case of autism, for instance, Bovell (2015) notes how 'even if the word "suffering" is not always used, much of the discourse in both academic and lay communities implicitly or explicitly relates to the relationship between autism and suffering or autism and reduced wellbeing, relative to a neurotypical "norm"' (p. 265, also see, e.g. Barnbaum, 2008; Furman & Tuminello, 2015). More broadly, Bickenback et al. (2014) note an 'engrained philosophical tradition' that holds that cognitive disability as such undermines 'the possibility of living the good human life' (p. 4).

My interest here is how this traditional depiction of cognitive disability is challenged by both the emerging neurodiversity paradigm, and by alternative models of disability available to neurodiversity proponents. For this chapter, 'neurodiversity' will simply refer to the brute fact of neurocognitive variation among the human species. By contrast, the 'neurodiversity paradigm' will be taken to refer to the theoretical and ideological shift towards reframing those who fall outside neurocognitive norms as 'neuro-minorities' marginalised by a 'neuronormative' organisation of society in favour of the 'neurotypical', rather than as a matter of individual medical pathology. On this view, rather than there being a relatively restricted medical notion of normality, neurocognitive diversity itself is the norm, much as it is normal and healthy for an ecosystem to be biodiverse. This is relevant to thinking about wellbeing since, on neurodiversity framing, neurodivergent suffering is primarily (although likely not totally) a product of societal exclusion and marginalisation, instead of neurodivergence being inherently pathological, tragic, and at odds with living a good life. Those I will call 'neurodiversity proponents' will thus be defined here as those who – although they surely disagree on much else – share a basic commitment to opposing the medical pathologisation of cognitive disability, both philosophically and ideologically, on these specific grounds.[2]

The debate between neurodiversity proponents and medical model proponents is often carried out more at the ideological level. For instance, neurodiversity proponents find tragic representations of cognitive disability dehumanising, whereas critics of the neurodiversity perspective have suggested that the latter can lead to overlooking the harsh realities of living with cognitive disability (Chapman, 2019a). However, underlying such disagreements are (often tacit) commitments to different models of disability in the more restricted philosophical sense. Indeed, arguably, the legitimacy of which ideological representations are adopted rests, in large part, on which model is more accurate in the philosophical sense. For if, say, autism really *is* a medical pathology, it would make perfect sense to view it as individual bad luck and to try and 'cure' it. But if it is in fact a non-pathological, marginalised minority after all, then it would seem wrong to try and cure it, and right to focus on inclusion and acceptance, as neurodiversity proponents argue. So, disagreements at the ideological level often, in large part at least, come down to underlying philosophical commitments to different models of disability. And by the same token, which model turns out to be best placed to capture the nature of any given disability will thereby provide support for corresponding practices and representations at the ideological level.[3]

With this in mind, my focus in this chapter will be on neurodiversity in relation to three different models of disability (taking 'model' in the more restricted, philosophical sense). These are the medical model, the social model, and the value-neutral model. I shall first show how the medical model (or at least one leading version) is unsuitable for accurately framing cognitive disability. This is because it is unable to provide an objective account of function and dysfunction, which makes it open to undue pathologisation of minority modes of functioning. Second, I argue that the social model, which is typically favoured by neurodiversity proponents and further undermines the medical model, is nonetheless also unsatisfactory. This is because it gets caught between one of two untenable positions: either hardship-denying or implicitly committing to an untenable species norm. Third, I suggest that the value-neutral model avoids the issues I identify with the medical and social models. This is a recent and lesser-known model of disability proposed by the philosopher Elizabeth Barnes (2009, 2016). Barnes herself has so far limited the scope of the model to physical disability, but I show how it can help us clarify the relationship between neurodivergence, disability, and wellbeing more fruitfully than either previous model. Although I cannot offer a thorough defence of the value-neutral model as applied to neurodiversity in this short chapter, I have at least given some preliminary reasons for thinking that this model avoids the problems encountered by both the social and medical models, while both supporting the neurodiversity paradigm perspective and undermining the notion that cognitive disability is at odds with wellbeing or living a good life.

Neurodivergence and the biostatistical medical model

The most influential clinical definition of the medical model as applied to neurodivergence can be found in the DSM-5's definition of mental disorder.[4] According to the DSM-5's definition (American Psychiatric Association, 2013, p. 20) a mental disorder is a 'a syndrome characterised by clinically significant disturbance in an individual's cognition, emotion regulation, or behavior that reflects a dysfunction in the psychological, biological, or developmental processes underlying mental functioning'.

In other words, to be counted as a mental disorder, on this view, something has to be considered both dysfunctional *and* harmful (to a 'clinically significant' extent), in the sense that the harm is taken to stem from the dysfunction. However, this instantly raises two further theoretical questions. The first regarding what it means to be dysfunctional, and the second regarding how we should understand harm. I will focus more on the former in this section and will come back to the latter when I discuss the social model.

Notably, there is no single universally accepted way of understanding the concept of dysfunction in a medical sense. And influential manuals like the DSM-5 purport to remain theory-neutral as to what the term 'dysfunction' means. Saying that, among the most influential unified models of natural dysfunction and pathology comes Christopher Boorse's biostatistical theory (BST) of health (1975, 1976, 1997). Although Boorse's is not the only naturalist medical model of dysfunction (also see Wakefield, 1992), his is among the more nuanced and robust, and moreover it provides a helpful contrast with the neurodiversity perspective.

To account for the notions of health and pathology, Boorse (1975, 1976) initially adopts a goal-oriented systems account of functioning. On this account, function is defined by the causal role it plays in fulfilling the goals of a biological system (in this case, the system is the organism). Each human organ (e.g. the heart), for instance, seems to play a specific role in each human organism's functioning, and, ultimately, their biological goals of survival and reproduction. Thus, for instance, we can tell the function of the heart is to pump blood, since this is the role it plays within a wider nexus of organismic systems that together contribute to the overall goal of survival and reproduction. In turn, Boorse further proposes that we can measure whether any given organ, or subsystem, is functioning naturally or not by measuring it in relation to the standard functioning for what he calls the *reference class*. This consists in the general class of those who are naturally uniform with each other. More specifically, Boorse proposes that the reference class for any individual organism should consist in those with the same species, age, and sex, these being the most obvious naturally uniform groupings. Hence, the functioning of any given organism's systems and subsystems, should be measured against the biostatistical norms for those with the same species, age, and sex, (in other words, the reference class) in order to check whether they are functioning normally or not when compared to the biostatistical average of their.

This is the basis for Boorse's biostatistical theory of health, whereby health is simply normal species functioning, and dysfunction (or pathology, impairment, or disease) refers to statistically sub-optimal functioning of any given organ. This allows Boorse to justify widely held intuitions regarding what should count as healthy or not: for instance, that it is biologically normal (and hence natural) for a 60-year-old man to be going bald but a sign of dysfunction if a seven-year-old girl begins going bald. That is, we can tell this by assessing each in relation to the biostatistical average for their reference class (i.e. all members of the same species, age, and sex). By using the notion of the biostatistical norms of the reference class, Boorse (1997) thus posits that dysfunction and pathology are an 'objective matter, to be read off the biological facts of nature without need of value judgements' (p. 4). When it comes to cognitive disability, the status and severity of dysfunctional can be objectively verified by assessing whether any given individual's neurocognitive functioning is suboptimal in relation to the biostatistical norm of the reference class in terms of its casual contribution to propensity for survival and reproduction. Any cognitive trait that is helpful for survival and reproduction (e.g. social understanding, intelligence, etc.) that is suboptimal in relation to the average of the reference class is thus taken as an objective dysfunction on the biostatistical theory of health.

Problems with the biostatistical medical model

While Boorse's biostatistical model is among the most careful and robust naturalistic conceptions of health and pathology, it has faced many criticisms, some of which chime well with the shift to the neurodiversity framing. I will focus on two here, which I take to undermine the purported objectivity of the biomedical framing.

On the one hand, some commentators have questions Boorse's reliance on the notion of the reference class (i.e. the class of all organisms with the same species, age, and sex). For instance, as Kingma (2007) stresses, it is not clear that we should only rely on species, age, and sex for the reference class. Perhaps, for instance, we could add, say, sexuality or – neurodiversity proponents might add – neurotype, at least for those neurotypes that seem naturally uniform (e.g. Down syndrome). This is important because, depending on what we decided to include or not, the outcome in terms of who is considered pathological or not will shift. For instance, if we decided to measure the cognitive ability of people with Down syndrome in relation to the species-average as such, they will be considered dysfunctional on Boorse's account; but if we decide to include the neurotype as part of the reference class, then many will not, since the average will change. But who is to say what we include in the reference class or not? This, for Kingma, remains unclear, leaving some level of arbitrariness, not to mention the possibility of being guided by normative assumptions, when it comes to who we count as dysfunctional or not. After all, it is far from clear that Down syndrome is any less naturally uniform than, say, sex.[5]

This relates to another potential issue noted by Varga (2015, p. 151). This regards how, in Boorse's own words, he 'presupposes enough uniformity in the species to generate a statistically typical species design' (1977, p. 562). Not much has been made of the fact that his theory rests on this presupposition in the academic literature (although see Amundson, 2000), but it is worth noting that Varga's point precisely fits with a core claim of neurodiversity proponents, which is to deny the assumption that there is such a level of uniformity in the neurocognitive functioning of our species. On the neurodiversity view, this would be like assuming dogs have enough uniformity across the species to assess the functions of different breeds in relation to the statistical norm, leading us to, say, consider all pugs physically dysfunctional and all Labradors cognitively dysfunctional – that is, it would be somewhat absurd and based on a category error (Chapman 2019a).

While many other criticisms of the medical model from a neurodiversity perspective focus on how harmful and dehumanising a deficit-based framing can be, here I have focused on how the medical model will turn out to be value laden in such a way that renders its purported objectivity suspect. Given what we have covered, it seems that, when challenged by the concept of neurodiversity, Boorse would need to beg the question in order to justify the pathologisation of neurodivergence. That is, to count neurodivergence as objectively dysfunctional, he needs to presuppose the precise points (i.e. the legitimacy of the species-standard norm, and the exclusion of neurotype from the reference class) that neurodiversity proponents contest. To my mind, the upshot of such worries is that, once challenged by the neurodiversity paradigm alternative, it would be hard to justify adopting such a medical model for cognitive disability. For it is not clear whether we can have a have an objective, value-free account of dysfunction as applied to the mind that is not at risk of undue pathologisation of minority.

Neurodivergence and the social model

In contrast to the medical model, neurodiversity proponents typically draw on the social model of disability (or some variation of it) in order to account for neurodivergent disablement and distress. This model mainly challenges the second aspect of the concept of mental disorder, which is harm rather than dysfunction. While I agree with social model proponents that the majority of problems faced in light of being cognitively disabled are political rather than medical,[6] I shall argue below that this model also faces problems of its own.

Traditionally, proponents of this model make a crucial distinction between impairment and disability, arguing that disability is caused not by impairment, but rather by how society fails to accommodate and accept impaired individuals (Oliver, 1990). For instance, a paraplegic person who uses a wheelchair is always impaired (this is counted as an objective fact), but they are only considered disabled when their impairment is not accommodated for, such as when there are only steps instead of ramps, or stigma regarding their impairment. This shifts the way disability is framed away from being seen as an individual medical issue, and towards it being a political issue. For on this model, the primary *cause* of disability is the way ableist societies are organised, rather than disability being framed as an individual matter.[7]

When it comes to neurodivergence, the social model allows us to reframe much of the disability and distress experienced by neurodivergent individuals as a political issue. For instance, the issues that come with the heightened sensory sensitivity often found among the autistic population can be framed as problems with disabling sensory environments, rather than stemming from an intrinsically bad processing style. In line with this, there is now lots of research showing how neurominorities can be thought of as being disabled due to attitudes and structures that fail to accommodate their minority cognitive styles (see, e.g. Milton, 2016; Roberston, 2010). I have argued for this position in detail elsewhere, with a particular focus on schizophrenia and what psychiatrists (problematically) call 'severe' autism (Chapman, 2019a). To the extent such accounts are convincing, the social model thus builds on the criticisms of the concept of dysfunction that we have already covered, in challenging the notion that those ways of functioning that fall below the species-norm are harmful.

Problems with the social model

The insights from social model framings provide further reason to be wary of the medical model, and the social model may also help avoid the risk of undue pathologisation I associated with Boorse's approach. To the extent this is convincing, the social model may allow us to move away from the tragedy narratives surrounding disability that so many disabled persons find harmful and dehumanising. When it comes to diminished wellbeing and flourishing, at least, the social model indicates that this may be a contingent product of society rather than being a necessary part of disabled life. Nonetheless, and despite my continued preference for the social model over the medical model, I contend that the neurodiversity proponent who also wants to use the social model will get caught between one of two untenable positions.

The first regards how the very concept of impairment (or similar alternative like 'disadvantage' or 'dysfunction') has both internal issues *and* clashes, conceptually, with the notion that diversity itself is the norm (i.e. one of the central commitments of the neurodiversity paradigm). The core reason for this is that to be considered impaired, you must be impaired in relation to something that is considered unimpaired. But as Barnes (2016, Ch. 2) notes, when we ask what this means, and who fits within which category, it is hard to see how we can avoid falling back into some notion of 'normal' functioning, and in turn to something (at best) like Boorse's biostatistical conception of normal function and dysfunction.[8] In fact, then, the social model does not wholly reject the naturalist reliance on a species standard-norm; rather, it only makes a different causal claim, that disability is caused by society more centrally than impairment.

The untenability of this position lies in two related issues. First, because the concept of impairment is so similar to the notion of natural dysfunction, then all the conceptual issues associated with the naturalist medical model re-emerge for the social model proponent. So anyone wanting to hold on to the concept of impairment (on whatever variation of the social model) will have to deal with those same issues (e.g. regarding making seemingly arbitrary choices as to what factors to include in the reference class, and so forth). In other words, they will encounter the same kinds of issues that I have noted with Boorse's biostatistical model. The second issue that arises here, at least for neurodiversity paradigm proponents, is that the neurodiversity paradigm rejection of the species-norm directly clashes with the concept of impairment: for the latter precisely presupposes that the species-norm is a legitimate standpoint from which to judge functioning levels, whereas the latter denies it. Put another way, once the shift to the neurodiversity paradigm is made, then the theoretical underpinning for the notion of (cognitive) impairment, which is a crucial part of the social model, has been lost. For holding diversity itself to be the norm does not seem compatible with holding the average to be the norm. This means that the neurodiversity paradigm is at odds with the concept of impairment as relied on for the social model.

Is the social model salvageable for the neurodiversity proponent? At this point a defender might reply that we should replace the concept of impairment with something neutral that does not presuppose a species norm. For instance, a neutral term like 'variation', as has been proposed by Scotch and Schriner (1997) might be taken to work. But this leads to the second untenable position. For if this route is taken, then it is unclear that the neurodiversity conception can account for the complexity of human cognitive disability. There are some potential intersections of cognitive variations that would, as far as can be reasonably inferred, make life a lot harder in certain ways regardless of how well supported they are. It would seem wrong to frame this as mere 'variation', which is a neutral term. The risk here is that slightly adapting the social model might end up with what we might call 'hardship denying' – in short, overlooking or bright-siding genuine issues that do primarily seem to stem from cognitive traits, and which cannot be easily reduced to a matter of marginalisation and oppression (this is a common critique of the social model in the disability studies literature, see, e.g. Shakespeare & Watson, 2001).

Either way, then, the social model seems to lead to a problem, especially for neurodiversity proponents. On the one hand, it could have room to admit the hardships that accompany disability, but in doing so falls back on a species-standard norm (by using notions like 'impairment') that is not just untenable but also contradicts the very notion that diversity itself is the norm; *or* it avoids these issues, but in doing so falls into hardship denying (i.e. denying the hardships that comes with certain cognitive disabilities). Overall, then, while the social model helps undermine and correct certain issues encountered by the medical model, most notably the notion that cognitive disability is inherently harmful, it is not fully satisfactory as a replacement.

The value-neutral model

I propose that we can turn to Barnes' (2009, 2016) value-neutral model of disability to help overcome these problems. Although Barnes herself developed this model primarily to account for physical disabilities, I will argue that it may be similarly useful for cognitive disability, allowing us to avoid the issues just noted with the social model while keeping its most important insights.

The first thing to note is that Barnes rejects the distinction between impairment and disability, mainly due to metaphysical problems the notion of impairment encounters, such as those I have already noted. Hence, Barnes just uses the term 'disability', and opts for a what she characterises as a 'moderate social constructivist' (2016, p. 38) view, whereby 'disability is socially constructed' to an extent, but also 'whether you are disabled is in part determined by what your body is objectively like' (ibid.). These are then two intertwined aspects of disability, rather than being neatly divided into objective impairment and societal disablement. The second thing to note is that 'value neutral' refers to being neutral in respect to its effects on wellbeing only. In contrast to the medical and social models, Barnes thus makes an *explicit* shift of focus to the importance of wellbeing. The point here is to clarify the core issue when it comes to models of disability, which regards, for her, whether disability is the kind of thing we should expect to reduce wellbeing even in a hypothetical world without ableism. If it is, then disability really is a 'bad difference', as the medical model indicates; but if it isn't, then this is more in line with the view that disabled people are a political minority but not necessarily medically pathological.[9]

Given these preliminaries, Barnes (2016, 80–82) proposes a crucial distinction between global and local wellbeing. Local wellbeing refers to wellbeing in some specific sense and at some specific time. By contrast, global wellbeing refers to wellness *on the whole*, or all things considered. This distinction is important because being locally bad does not necessitate being globally bad. For instance, getting up very early to run before work could be locally bad (e.g. if you hate having to wake up early), but globally good (if it makes you healthier, feel good about yourself, and so forth) (p. 81). With this distinction in mind, Barnes opens up the possibility of taking things to be simultaneously different in terms of their value (good, bad, or neutral) at different levels without any contradiction. Based on this, Barnes argues, although disabilities do often come with local bads – that is, things that can make life locally harder – this does not necessitate that disability is the kind of thing that is inherently bad for global wellbeing. In fact, she notes (2017, 71–72) lots of empirical research indicates that physical disability precisely does not tend to make global wellbeing worse, even though stigma and marginalisation do (also see Carel (2016, 130–150) for a recent review and analysis). For Barnes, this is corroborated by the testimonies of numerous disabled individuals who explain how their disabilities add value to their lives as well as making life harder (2009, 341–342). With this in mind, Barnes (2016, p. 98) concludes that disability:

> may be good for you, it may be bad for you, it may be utterly indifferent for you – depending what it is combined with. But disability [as such] isn't itself something that's bad. It is, like many other minority features, neutral with respect to well-being.

In other words, although she does of course not deny that disabled individuals can be deeply unhappy, she establishes that disability is both value neutral *as such*, in terms of its intrinsic effects on global wellbeing, and yet simultaneously the kind of thing that can and often does make life harder locally. Hence, Barnes avoids hardship-denying while also denying that disability is tragic and objectively bad.

Neurodivergence and the value-neutral model

Barnes limits her argument to physical disabilities. Nonetheless, like the physically disabled, many neurodivergent individuals also testify that they are happy or capable of flourishing. To quote a few:

> My [autistic] personhood is intact. My selfhood is undamaged. I find great value and meaning in my life, and I have no wish to be cured of being myself.
> (Sinclair, 1993, p. 302)

> In spite of being 'disabled', I have managed to adapt quite well, and build a rich, full life (and I am far from unique in that regard).
> (Schneider, 1999, 10–11)

> I am a man with Down syndrome and my life is worth living. In fact, I have a great life!
> (Stevens, quoted in Friedersdorf, 2017, n.p.)

One view would be that these are merely anomalies – the exceptions that prove the rule that in general cognitive disability brings a propensity towards low levels of wellbeing. But I suggest that we should not expect cognitive disability to be inherently associated with low global wellbeing and more than neurotypicality is at all. There are two key points in support of this.

The first key point is that, like physical disabilities on Barnes' account, cognitive disabilities can be good or bad depending on the *external* context (much as social model proponents argue, albeit with more of a focus on wellbeing). Take the example of autism. It is well known that wellbeing is generally lower among the autistic population than the non-autistic population. On the medical model, low wellbeing is often taken to stem from being autistic (although it does not deny that social factors may contribute too). Nonetheless, a study by Renty and Royers (2006) found that wellbeing among the autistic population was predicted not by how 'severe' the level of impairment was, but rather by whether they felt well supported (in other words, it was determined by the external context, not where on the spectrum any individual was). A different example is that of schizophrenic hallucinations. While these are typically thought to be intrinsically harmful, a recent anthropological study (Luhrman et al. 2014) found that whether or not they were experienced as harmful or not was largely determined by the culture of the individual experiencing them. They found that in North America, hearing voices was generally experienced as distressful, whereas in Ghana and India such voices were generally experienced as benign and playful. So this cognitive feature will only lead to reduced wellbeing in certain social contexts. What such findings suggest is that the notion that reduced wellbeing among the cognitively disabled stems from the underlying differences is not obviously always founded – there is good initial reason to think that it is determined in significant part by external factors. In other words, even if these cognitive disabilities make life inherently harder in certain ways, there is good reason to think that this means they will necessarily reduce global wellbeing. In the words of Naoki Higashida, himself classified as being 'severely' autistic on the medial framing, 'functioning in our society is difficult for neuro-atypicals, but encountering difficulties is not the same thing as being unhappy' (2017, p. 261).

The second key point is that specific cognitive traits can also be good or bad *in relation* to other *internal* factors, these being one's other cognitive traits, and ultimately, one's overall cognitive makeup. This is important here insofar as what is often be thought of as simply bad might actually only be so in combination with other internal traits. Consider a study by Palmer, Martin, Depp, Glorioso, and Jeste (2014) on wellbeing among the schizophrenic population. This study found not only that many schizophrenic patients were happy, but moreover that happiness levels were primarily correlated with psychological traits unrelated to schizophrenia as such. For instance, propensity towards optimism was strongly correlated with happiness, even though there was no correlation between wellbeing levels of the cognitive traits definitive of schizophrenia. Here, then, it seems that whether schizophrenic cognitive disabilities manifest as good or bad overall depends on other internal factors (i.e. the rest of each individuals cognitive profile, including how optimistic they were) more than the disabilities themselves. So whether the cognitive disabilities associated with schizophrenia manifested as bad came down to other internal factors.[10] A different example comes from a study by Skotko, Levine, & Goldstein (2011) on levels of wellbeing among the population with Down syndrome. They found not just that they were mostly happy, but rather that their levels of happiness were strikingly high. What this suggests is that the person with Down syndrome may have a cognitive profile that, on the whole, is better geared towards high levels of wellbeing than those who fall within the normal range of cognitive functioning (Skotko et al., 2011). In this case the various cognitive disabilities associated with Down syndrome thus seem to contribute towards happiness rather than diminish it. Although this is of course not the case with all cognitive disabilities, here we see that in some cases cognitive disability contributes towards wellbeing. Overall, then, whether a cognitive disability (or ability) is good or bad will not be an essential property of the cognitive trait or traits definitive of that disability, but will rely in large part on the other cognitive traits they happen to be combined with.

Taken together, the upshot of these two points is that any cognitive ability or disability that is actualised in a way experienced as a negative for one person might be actualised in a way that is experienced as a positive for another, depending on both their overall cognitive style (i.e. which other cognitive traits it is combined with) *and* on the context they are in. This gives good preliminary reason to think that cognitive disability may be very similar to physical disability as Barnes (2016) portrays it, in so far as it 'can sometimes be bad for you – depending on what (intrinsic or extrinsic) factors it is combined with. But it can also, in different combination, be good for you' (p. 88). This model allows us to coherently hold that, although cognitive disability surely does often make life intrinsically harder in various respects, there is simultaneously good reason to think that it is value neutral at the global level. For it is not good or bad in itself; rather, it only becomes one or the other when combined with other factors.

Complex cases

I have argued that there is good initial reason to think that the value-neutral model is a preferable theoretical model for the neurodiversity paradigm, since it avoids the theoretical issues I have noted with the social and medical models. Moreover, while the issue of wellbeing is usually tacitly there in the background of these debates, bringing it to the fore may help provide a more direct and challenge to the depiction of cognitive disability as tragic and at odds with the good human life. The value-neutral model may therefore have potential to provide a more nuanced and realistic account of neurodivergent disablement, while retaining, and in some ways supplementing, the radical political utility of the social model. Nonetheless, there are some potential criticisms and worries that I have not yet discussed. I will end by discussing what I take to be the most obvious criticism, although I suspect there are others that will need to be dealt with elsewhere.[11]

The worry I take to be most obvious regards how there may be some complex cases where there is, seemingly, more reason to think that reduced wellbeing is likely to have been caused by some combination of cognitive disabilities. I have in mind here how critics of the neurodiversity paradigm typically offer examples of specific multiply disabled individuals who do seem more likely to have hindered quality of life regardless of their support levels. In reply, let's assume for a moment that there are some cognitively disabled persons who do have an inherent disposition towards being very unhappy, and that this does seemingly stem from their overall cognitive makeup. Even if we do allow this assumption, does this show that there is anything about cognitive disability as such? I would suggest not.

The first thing to note here is that the value-neutral model would predict that there would be some such individuals. For even if cognitive disability is value neutral in itself, it can be bad in combination with other cognitive traits (e.g. as we saw in the combination of schizophrenic traits combined with low levels of optimism). It may be that some specific combinations of traits are bad in combination even if they are not singularly. Second, it is also worth considering that there are equally many people who are cognitively able, and who fall within what Boorse would consider the normal range of cognitive functioning, who likewise seem to have such a disposition towards low levels of wellbeing due to the idiosyncratic combination of normal-range cognitive traits each has. At least, it seems to me that there are a great many deeply unhappy neurotypicals, and although in many cases this will surely be due to contingent social factors, in others it will be more closely related to their underlying neurocognitive makeup (which will be unique in each case, even though they all fall within a statistically normal range of functioning).[12] Put another way, my point is this: when we consider that each individual brain and correlating overall cognitive style is unique (even among those of any given neurotype), and while many individuals may very well have overall neurocognitive makeups that do indeed seem to tend towards low levels of wellbeing, I am unconvinced that there is anything about cognitive disability as such that makes this so, especially bearing in mind that many neurotypicals have similar natural dispositions towards unhappiness. Low wellbeing seems to be something that some humans have a neurologically based propensity towards, rather than this being cognitive disability specific. There is reason to think that cognitive disability, in itself, thus remains value neutral, or at least no less so than cognitive ability, even if there were some instances where cognitively disabled individuals did have an inherent propensity towards low wellbeing.[13]

Of course, I do not deny that there is a wide-spread *intuition* that cognitively disabled lives are, qua cognitive disability, inherently disposed to be worse off in some other, more objective sense. Because of this intuition, here the medical model proponent might make the further (problematical) claim that the happy neurodivergent is like the 'happy slave'[14] who turns down the offer of freedom since they are genuinely (subjectively) happy, albeit only because they do not fully understand how (objectively) bad their situation is. But this is the very intuition that I have argued is unjustified given the evidence presented here. So whoever wants to argue that the happy neurodivergent is still objectively worse off will need to find some independent argument that does not implicitly rely on an assumed species-standard functional norm or other neuronormative assumptions or intuitions about what the good life consists in.

Acknowledgements

I would like to thank the editors for their hard work editing this volume and Havi Carel for helpful feedback on a previous draft.

Notes

1 Here I will use 'cognitive disability' as an umbrella term to include biologically based developmental, sensory, and learning disabilities. Although the term 'disability' is often used as a medical term, I use it without assuming any necessary connection.

2 The concept of neurodiversity was coined by Blume (1998) and Singer (1999). Here, I follow Walker's (2014) influential framework, which can be accessed online. To see my own clarification and defence of this framework, see (Chapman, 2019a).

3 Arguably, there is no *necessary* connection between what I have framed as the philosophical and ideological variants of the medical model. But in practice they are so deeply intertwined that this seems a merely technical point.

4 Although, there are instances of neurodivergence that are not in the DSM-5, for instance Down syndrome; and there are also diagnostic categories listed in the DSM-5 that I am happy to accept are genuine mental disorders, for instance, post-traumatic stress disorder, major depressive disorder, and the various anxiety disorders. For I take neurodiversity to be mainly concerned with cognitive, sensory, and developmental variation rather than psychological responses to trauma and stress.

5 There does seem to be a case for adding Down syndrome on the Bornean framework insofar as this neurotype does seem like a naturally uniform kind. However, there may be less reason when it comes to other neurotypes, such as neurotypicals or autistics, since these are heterogenous groupings determined by society rather than being natural kinds (on this see Chapman, in press).

6 I have defended the social model as applied to various forms of neurodivergence elsewhere (Chapman, 2019a, 2019b, in press). I take the arguments I made there to be supplemented by those I will make here, rather than being at odds with them.

7 This, at least, is how the traditional British social model frames it. It is important to note that there have been updated variations on the medical model, for instance the Nordic relational model, which takes both impairment and societal disablement to be casually relevant. For the sake of clarity, I shall stick with the British model, but it should also be noted that the criticism I make below would apply equally to any variant of the social model (such as the Nordic model) that still retains the concept of impairment.

8 'Dysfunction' and 'impairment' are often used interchangeably, there being no clear conceptual difference between the two.

9 Although, depending on how we understand the social model (and although it is used to combat the personal tragedy view), it could be taken to indicate that impairment

would still be bad in this sense even in a world without ableism, or that disability would have no negative effect on wellbeing. For the distinction between disability and impairment leaves open whether impairment can be inherently bad. But either way, it would encounter one of the two problems noted above (that is, on the former reading, it would encounter all the issues associated with the naturalist model, whereas on the latter, it may end up impairment-denying).

10 An interesting practical implication of this is that, if a hypothetical scientist with advanced neurology-editing techniques wanted to edit an unhappy schizophrenic individual's wellbeing levels, then it seems that slightly tweaking their already normal-range traits (e.g. levels of optimism) could very well be more fruitful than minimising the cognitive traits associated with schizophrenia

11 For a nuanced discussion of some potential issues, see (Barnes, in press).

12 It is interesting that high intelligence has been associated with a risk of depression and other similar issues (Karpinski, Kinase Kolb, Tetreault, & Borowski, 2018).

13 It is worth saying here that, in such cases where an individual had an inherent disposition towards low wellbeing, it may be right to treat this as a medical issue (at least in cases where they feel they would benefit from doing so). But this would be so for the depression disposed neurotypical as much as the neurodivergent, and so is again nothing to do with cognitive disability as such. Indeed, although I do not want to go into this here, this may help us distinguish between genuine mental illness, and disabled (but not pathological) forms of neurodivergence. On the view I propose here, it may be that both some neurotypicals and some neurodivergent individuals are mentally ill, in so far as their individual cognitive styles do hinder their flourishing, but that there is no reason to think that neurodivergent disability, or level of deviation from the norm, should be more closely associated with mental illness than neurotypicality is.

14 This is the hypothetical example often given in the literature, but it is important to stress that the example itself is based on a deeply problematic myth.

References

American Psychiatric Association (2013). *Diagnostic and statistical manual of mental disorders* (5th ed.). Washington, DC: American Psychiatric Publishing.

Amundson, R. (2000). Against normal function. *Studies in History and Philosophy of Science Part C: Studies in History and Philosophy of Biological and Biomedical Sciences*, *31*(1), 33–53.

Barnbaum, D. R. (2008). *The ethics of autism: Among them but not of them*. Bloomington, IN: Indiana University Press.

Barnes, E. (2009). Disability, minority, and difference. *Journal of Applied Philosophy*, *26*(4), 337–355.

Barnes, E. (2016). *The minority body: A theory of disability*. Oxford: Oxford University Press.

Barnes, E. (in press). Replies to Dougherty, Kittay, and Silvers. *Philosophy and Phenomenological Research*, *100*, 232–243.

Birchenback, J. E., Felder, B., & Schmitz, B. (2014). Introduction. In J. E. Bickenbach, F. Felder & B. Schmitz (Eds.), *Disability and the Good Human Life* (pp. 1–18). Cambridge: Cambridge University Press.

Blume, H (1998). Neurodiversity: on the neurological underpinnings of geekdom. *Atlantic*. Available at www.theatlantic.com/magazine/archive/1998/09/neurodiversity/305909/

Boorse, C. (1975). On the distinction between disease and illness. *Philosophy & Public Affairs*, *5*(1), 49–68.

Boorse, C. (1976). What a theory of mental health should be. *Journal of Social Behaviour*, *6*, 61–84.

Boorse, C. (1977). Health as a theoretical concept. *Philosophy of Science*, *44*, 542–573.

Boorse, C. (1997). A rebuttal on health. In J. M. Humber & R. F. Almeder (Eds.), *What is disease?* Totowa, NJ; Humana Press.

Bovell, V. (2015). Is the prevention and/or cure of autism a morally legitimate quest? (Unpublished doctoral thesis). University of Oxford.

Carel, H. (2016). *Phenomenology of illness*. Oxford: Oxford University Press.

Carlson, L. (2010). *The faces of intellectual disability*. Bloomington, IN: Indiana University Press.

Chapman, R. (2019a). Neurodiversity and its discontents: Autism, schizophrenia, and the social Model. In S. Tekin, & R. Bluhm (Eds.), *The Bloomsbury companion to the philosophy of psychiatry* (pp. 371–389). London: Bloomsbury.

Chapman, R. (2019b). Autism as a form of life: Wittgenstein and the psychological coherence of autism. *Metaphilosophy*, *50*(4), 421–440.

Chapman, R. (in press). The reality of autism: on the metaphysics of disorder and diversity. *Philosophical Psychology*.

Friedersdorf, C. (2017). I am a man with Down syndrome and my life is worth living. *The Atlantic*. URL: (Retrieved December 2019) www.theatlantic.com/politics/archive/2017/10/i-am-a-man-with-down-syndrome-and-my-life-is-worth-living/544325/]

Furman, T. M., & Tuminello, A. (2015). Aristotle, autism, and applied behaviour analysis. *Philosophy, Psychiatry, & Psychology*, *22*(4), 253–262.

Higashida, N. (2017). *Fall down seven times: Get up eight. A young man's voice from the silence of autism*. London: Hodder & Stoughton.

Karpinski, R. I., Kinase Kolb, A. M., Tetreault, N. A., & Borowski, T. B. (2018). High intelligence: A risk factor for psychological and physiological overexcitabilities. *Intelligence*, *66*, 8–23.

Kingma, E. (2007). What is it to be healthy? *Analysis*, *67*(294), 128–133.

Luhrmann, T. M. R., Padmavati, H., & Tharoor, A. O. (2014). Differences in voice hearing experiences of people with psychosis in the USA, India and Ghana: Interview-based study. *The British Journal of Psychiatry*, *206*, 41–44.

Milton, D. E. M. (2016). Disposable dispositions: reflections upon the work of Iris Marion Young in relation to the social oppression of autistic people. *Disability and Society*, *36*(10), 1403-1407.

Oliver, M. (1990). *The politics of disablement*. Basingstoke: MacMillan.

Palmer, B. W., Martin, A. S., Depp, C., Glorioso, D. K., & Jeste, D. V. (2014). Wellness within illness: happiness in schizophrenia. *Schizophrenia Research*, *159*, 151–156.

Renty J. O., & Roeyers H. (2006). Quality of life in high-functioning adults with autism spectrum disorder: the predictive value of disability and support characteristics. *Autism*, *10*(5), 511–524.

Robertson, S. M. (2010). Neurodiversity, quality of life, and autistic adults: Shifting research and professional focuses onto real-life challenges. *Disability Studies Quarterly*, *30*(1)

Savulescu, J., & Kahane, G. (2011). Disability: A welfarist approach. *Clinical Ethics*, *6*(1), 45–51.

Scotch, R. K., & Schriner, K. (1997). Disability as human variation: Implications for policy. *The Annals of the American Academy of Political and Social Science*, *549*, 148–159.

Schneider, E. (1999). *Living the good life with autism*. London: Jessica Kingsley.

Shakespeare, T., & Watson, N. (2001). The social model of disability: an outdated ideology? In S. Barnartt & B. Altman (Eds.), *Exploring theories and expanding methodologies: Where we are and where we need to go (Research in social science and disability, Vol. 2)* (pp. 9–67). Bingley: Emerald Group Publishing Limited, Bingley, pp. 9–28

Sinclair, J. (1993). Don't mourn for us. *Our Voice*, 1(3). URL: (Retrieved December 2019) www.autreat.com/dont_mourn.html

Singer, J. (1999). Why can't you be normal for once in your life? From a 'problem with no name' to the emergence of a new category of difference. In M. Corker & S. French (Eds.), *Disability discourse* (pp. 59–67). Buckingham: Open UP.

Skotko, B. G., Levine, S. P., & Goldstein, R. (2011). Self-perceptions from people with Down syndrome. *American Journal of Medical Genetics Part A, 155*, 2360–2369.

Varga, S. (2015). *Naturalism, interpretation, and mental disorder*. Oxford: Oxford University Press.

Wakefield, J. (1992). The concept of mental disorder - on the boundary between biological facts and social value. *American Psychologist, 47*, 373-388.

Walker, N. (2014). Neurodiversity: Some basic terms & definitions. *Neurocosmopolitanism*. URL: (Retrieved December 2019) https://neurocosmopolitanism.com/neurodiversity-some-basic-terms-definitions/

Chapter 5

Neurodiversity in a neurotypical world

An enactive framework for investigating autism and social institutions

Alan Jurgens

In this chapter I argue enactivist accounts of the role social practices and institutions play in shaping cognition and identity reinforce the neurodiversity paradigm's focus on shifting the attention of interventions away from neurodivergent individuals and towards society.[1] Though I argue the enactive framework is especially suited for investigating and explaining neurodiversity, enactivist accounts have yet to engage with the issues raised by the social model approach to neurodiversity (Chown & Beavan, 2011) and two explanatory theories of autism, monotropism (Murray, Lesser, & Lawson, 2005) and the double empathy hypothesis (Milton, 2012a). While I draw on the established enactive accounts of De Jaegher (2013a) and Krueger and Maiese (2019) to show the advantages of enactivism, their work remains problematically committed to a medical model approach for explaining neurodiversity and proposing intervention strategies.

A medical model approach views physical or cognitive differences as disabilities, which are functional deficits that an individual either has or does not have. As the medical model approach views cognitive differences found in neurodiversity as deficits to be corrected, the intervention strategies on this approach are directed primarily at the neurodivergent. This can be seen in enactivist accounts, for example, where De Jaegher (2013a, p. 10) claims autistic self-stimulatory behaviour is something that we need to 'find suitable ways to deal with ... even to the point of converting them into acceptable activities or extinguishing them'. Similarly, Krueger and Maiese (2019) propose co-constructing inclusive music therapy institutions for improving autistic individuals' coordination and rhythm skills through musical practice. Both of these accounts' approaches to intervention strategies follow the medical model in seeing autistic differences as faults to be corrected primarily by the autistic individual him/herself. However, medical model attitudes have been challenged by the neurodiversity paradigm on the grounds that they are partly responsible for the creation of systemic barriers and negative stigmas the neurodivergent regularly face (Chown & Beavan, 2011).

In line with this challenge, the social model alternatively approaches neurodivergent differences by examining how society's normative expectations define their differences, and potentially cause further psychological 'disability' via 'a culture and ideology of "normalcy" ' (Milton, 2012b, p. 10). As the social model examines neurodiversity from a social perspective, the model explicitly rejects the idea that the cognitive differences of the neurodivergent are faults of the individual to be corrected. Additionally, social model approaches shift the focus of intervention strategies for improving neurodivergent capabilities and wellbeing away from putting the onus for change on the neurodivergent individual him/herself. Instead, social model approaches to intervention strategies focus on improving neurotypical social practices and institutions to be more inclusive and accommodating for the neurodivergent.

Despite previous enactivist commitments to a medical model approach, enactivism is naturally suited to a social model approach. This is because enactivism focuses its investigations of cognition through the concept of intersubjectivity. At the core of the concept of intersubjectivity is the claim that the structure of human experience and cognition always involves a relation to the world and to others, which constitutively shapes cognition. This is understood as how one experiences the other as someone to whom things matter, who one can connect with through feeling, and who is a distinct other with their own perspective (De Jaegher, 2018). Intersubjectivity is used to examine how relating and interactions create and transformation meaning, and through which individuals' perspectives on the world, each other, and themselves evolve.

Intersubjectivity is further conceptualised by examining the ways a particular individual's embodiment and embeddedness constitute the individual's cognition and experiences in the world. The terms embodiment and embeddedness denote the claims respectively that an individual's physical body beyond her brain, and her socio-material environment beyond her body, play a significant constitutive, not just causal, role for her cognitive processes. The claim that cognition is constituted via embodiment and embeddedness is to be understood in the sense that these aspects not only shape cognition but they are also necessary for it to arise in the first place. For enactivism, an individual's particular embodiment and embeddedness constitutively shape, via experiences in the socio-material world, cognition.

On this framework, as an individual's embodiment and embedded environment partly constitutes her cognition, the significance that the world has for her is not simply pre-given, but it is enacted through her interactions with others and her socio-material environment. As a result, enactivism's emphasis on intersubjectivity brings to the forefront of investigations of cognition first-person perspectives that in turn are able to help explain the normative effects social practices and institutions have on individuals' cognition and personal identity. For enactivism, intersubjectivity is used to help explain how individuals' personal identities are constituted through their relationships to the world in regard to self-image, self-esteem, individuality, and social position within society.

The focus enactivism places on examining intersubjective effects on identity construction aligns with the neurodiversity paradigm's claim for a need for a shift in terminology relating to autism. First, advocates of the paradigm claim terms such as 'Autism Spectrum Disorder' should be revised to reflect the preference of autistic self-advocates. In the autism community being autistic is, for many, central to their identity formation, and as such, it is not something suffered from. Rather, for these autistic individuals, being autistic is an integral part of their personal identity (Fenton & Krahn, 2007). Autistic self-advocates claim this shift in terminology helps to promote a positive self-image, while also countering negative bias autistic individuals experience in interactions with neurotypical individuals. Second, it is important to recognise the non-universality of autism, and that every autistic individual has a range of strengths and weaknesses just as every neurotypical individual does. In usage here, the term 'autism' simply refers to the whole range of the spectrum, whereas the term 'autistic individual' highlights the particularised nature of each individual's cognitive skills and behavioural habits.

While enactivism is a framework for explaining cognition, in order to examine the differences between autistic individuals and neurotypical individuals, in this chapter I integrate enactivism with the monotropic theory of autism. Monotropism claims 'atypical patterns of attention' are a core inherent feature of autism from which many of the notable social differences arise (Murray et al., 2005, p. 139). According to this account, there is a difference between the monotropic attention patterns of autistic individuals that involves having few, but intensely focused, attentional interests, and polytropic attention patterns of neurotypical individuals that involves having many, but less focused, attention patterns. Monotropism claims these differences in attention patterns can explain many of the other cognitive and behavioural differences documented between autistic individuals and neurotypical individuals. Some of these differences that are examined here are differences in: language development; sensorimotor synchronisation and coordination in social interactions; sensory differences in regard to hyper/hyposensitivity; and having a higher propensity for self-stimulatory behaviours.

In order to understand how these differences affect interactions autistic individuals have with neurotypical individuals, Milton's (2012a) double empathy hypothesis proposes there is a double empathy problem. Milton claims that since autistic and non-autistic individuals have 'different dispositional outlooks and personal conceptual understandings', when interacting with one another both groups are more susceptible to frequently misunderstanding one another (p. 884). It's a 'double problem' as the difficulty in understanding one another is bi-directional, arising from differences between the neurotypical individual and the autistic individual. As we live in neurotypical societies, societal institutions are structured in neurotypical-friendly ways and most of our interactions with these institutions are via individuals who are members, or representatives, of the institutions. For this reason, the double empathy hypothesis is also relevant for examining autistic individuals' interactions with and relationships to neurotypical social institutions.

In regard to this, Fenton and Krahn (2007, p. 1) identify that social institutions 'either expressly or inadvertently model a social hierarchy' in which the 'interests or needs of individuals are ranked relative to what is regarded as properly functioning cognitive capacities'. The neurodiversity paradigm is motivated in part by challenging this kind of social hierarchy found in institutions by reweighing the 'interests of minorities so that they receive just consideration with the analogous interests of those currently privileged by extant social institutions' (p. 1). Enactivism is well-suited for this task as it explains cognition in relation to social practices and institutions by (re)conceptualising individuals and their relationships to others, societal roles, and social interactions that constitute their cognition and identities (De Jaegher, 2013b).

The goal of this chapter is to show that enactivism as a general framework for explaining cognition is especially appropriate both for integration with the monotropism theory and the double empathy hypothesis and for explaining autism. This is because enactivism offers a holistic framework for examining the specific contributions internal (embodied) and external (embedded) factors play in shaping cognition. Though the social model approach is committed to highlighting the influence social environments have on shaping neurodivergent cognition, it is not in itself a framework for a comprehensive explanation of neurodiversity. Enactivism, as a philosophical and scientific framework for explaining cognition in general, already shares this core commitment of the social model through its focus on intersubjectivity. As such, enactivism is especially suited for developing a systematic explanation of not only autism, but also alternative forms of neurodiversity, along the lines of the social model approach through its focus on how a neurodivergent individual's embodied and embedded differences affect his/her intersubjective relationships.

In order to demonstrate how enactivism can more fully embrace a social model account and be a framework for explaining neurodiversity, I begin in Section 2 by providing an overview of the enactive social cognition framework. In Section 3, enactivism is then integrated with the monotropism theory in order to show how the double empathy problem arises and to examine the relationships autistic individuals have with neurotypical social practices and institutions. Finally, the chapter concludes by briefly discussing two further research paths based on the preceding analysis for the development of interventions to improve the social cognitive skills and wellbeing of autistic individuals. The first research path focuses on the potential the enactive framework has for explaining and further developing the field of animal-assisted therapy for autistic individuals (Smith, 2018). The second path raises the question of what an enactive neurodiversity paradigm in education would look like.

Enactivism, social practices, and institutions

A core proposal at the centre of the enactive framework for explaining cognition is the claim that an individual's particular form of embodiment determines what stimuli in the environment the individual is sensitive and responsive to (Maiese, 2018). Essentially, this means that by examining the way an individual moves and perceives, her sensorimotor system, it is possible to understand how her cognitive capacities and processes function and develop. Enactivism claims an individual's embodiment and embeddedness partly constitute her cognition; and at the same time, her actions in her environment change (or enact) the environment to better suit her needs and purposes. Importantly, the term 'constitution' as it is used here, and throughout this chapter, should be taken as a species of causation, that is, constitutive causation. Constitution used in this enactive sense is meant to capture the bidirectional aspect of enactive relationships where there is continuous reciprocal causal influence between individuals and their environments (Jurgens & Kirchhoff, 2019).

In regard to the social realm, enactivism utilises the concept of intersubjectivity to explain cognition by examining the salience various aspects of the sociomaterial world have for an individual, and how the individual interacts with these worldly phenomena. The most basic form of intersubjectivity is primary intersubjectivity. Primary intersubjectivity is claimed to develop in the first year of life where infants become capable of imitating facial expressions, which provides them with a basic sense of familiarity with others (Fuchs, 2015). This results in infants being affected by others' expressive behaviour, entering into a relationship of shared bodily feelings and affects. The capacity for secondary intersubjectivity develops through experiences of joint attention, gaze-following, and pointing as infants begin to be able to refer explicitly to the shared social and material context. Through experiencing how others interact with the world, infants pragmatically learn the meaning of objects and how to use them. Finally, tertiary intersubjectivity develops when children both understand that others may have conflicting perspectives and become able to shift between their own perspectives and the perspectives of others.

A key aspect of all of these levels of intersubjectivity is social normativity, which plays a pivotal role in explaining how individuals form values, attitudes, desires, conceive of thoughts, and execute intentions through action (Maiese, 2018). Individuals pick up social norms via their interactions with others by the embodied behaviours others have adopted from their own experiences with social practices and institutions. This is because 'social institutions enhance specific patterns of thought, feeling, and behaviour by providing a normative framework that rewards, reinforces, or discourages' particular kinds of ways of thinking and behaving (p. 12). One of the best, and developmentally earliest examples of how intersubjectivity both encourages, or discourages, normative behaviour is the infant–caretaker dyad. This example also demonstrates how the dynamics of an interaction can be constitutively constrained by the socio-material environment, including the other interactor.

In this kind of interaction, body posture, expressive gesture, vocalisation, gaze following, and attention are essential to maintaining an ongoing and recurrent engagement between infant and caretaker (Jurgens & Kirchhoff, 2019). Still face experiments (Nagy, Pilling, Watt, Pal, & Orvos, 2017) utilising infant–caretaker dyads show the importance of ongoing and synchronous engagement by demonstrating what occurs when the synchrony formed by these kinds of behaviours breaks down. As the core features of this kind of interaction are the infant and caretaker recognising, attending to, and responding to each other's interaction, the still face experiment shows that when the caretaker suddenly adopts a still face and no longer interacts with the infant, the infant becomes noticeably discouraged and upset. At this time, infants 'withdraw from the interaction, avert their gaze, display negative affect, become increasingly distressed, start crying and smile less' than during the previous interactive engagement phase (Nagy et al., 2017, p. 2). Furthermore, even when the caretaker re-engages with the infant after the still face phase, there is a spillover effect where the infant will continue to avert its gaze, display distress, and generally will not re-engage with the caretaker to the same level as the initial interactive phase (Nagy et al., 2017). This shows beyond a mere causal effect that the caretaker's behaviour of adopting a still face, where attention is no longer being paid to the infant, constitutively affects both the infant's social cognition during the still face phase and interactions following this phase.

According to enactivism, what explains the still face experiment is primary intersubjectivity. However, without attending to the right aspects of the other's embodied behaviours, developing secondary and tertiary intersubjectivity and more sophisticated social skills becomes more difficult. This is partly due to the fact that in order to develop social skills the infant not only needs to attend to the right aspects of the other's body, but also attune to the other's rhythm of movements. It is from attending and attuning to others, through interactive social experiences in the world, that infants develop more sophisticated social cognitive skills. In this view, the interactive gestures the caretaker directs towards the infant has a constitutive effect on how the infant perceives, moves, and emotes.

As De Jaegher (2013a) points out, an attuned rhythm capacity determines, among other things, one's timing and coordination in interactions with others. Being able to coordinate with others' behaviours, gestures, and utterances makes it easier to fluidly develop new social capacities and skills. The particular bodily gestures and vocalisations the caretaker directs towards the infant, which are based on the caretaker's history of culturally acquired social practices, begins to enculturate the infant in these specific practices. Following from this, we can see that immediately from birth the infant is immersed in social practices, highlighting the deep significance intersubjectivity has on shaping embodied social cognitive habits and processes. It is from these kinds of interactions that the infant already begins to develop culturally specific social skills.

It is only through our interactions with, and attending to, others and the cultural practices they embody, that we are able to develop specific social skills for understanding others (De Jaegher, 2013a). These social skills are the embodied capacities that we develop to flexibly respond to the regularities, and irregularities, found in interactions with individuals. These skills develop through a history of interactions with individuals and are partly constructed by the norms and practices of the society, or societies, to which one is exposed. Thus, in order to understand how social skills develop, enactivism claims we need to examine the constitutive effects social practices and institutions have on embodied social habits.

Social practices and institutions take many different forms across different cultures and times as they are the bonds that hold societies together by providing normative frameworks for interactions. Enactivism claims individuals' identities, cognitive processes, and social skills are constituted by 'social and cultural laws, regulations and norms' (De Jaegher, 2013b, p. 23). Some examples of social practices that Krueger and Maiese (2019) identify are: lining up in queues; pausing in conversations to allow the other to finish a thought; or expressing disapproval with a well-timed eyebrow raise. Examples of institutions can range from the concept and practices of families to institutions that rely on a multitude of other institutions, such as international law that requires multiple other intersecting institutions like justice systems, governments, national boarders, etc.

Importantly, De Jaegher (2013b, p. 23) claims that a central feature of all institutions is that

> interactions with institutions often happen at the 'face-to-face' level.' It is through our interactions with someone who represents an institution that we ultimately are influenced by the institution and its particular set of approved and regulated practices. Through our interactions with representatives of institutions we begin to embody 'certain models of expectancy [that] come to be established, and the patterns, which over time emerge from these practices, guide perception as well as action.
> (Roepstorff, Niewöhner, & Beck, 2010, p. 1056; cited in Krueger & Maiese, 2019, p. 21)

However, this doesn't necessarily mean that the person has to be an official representative of the institution. We don't need to interact with a police officer, a lawyer or judge to interact with someone who represents the institution of the law. A parent or teacher can also serve as a representative of the law simply by displaying acceptable lawful behaviour or discouraging unlawful behaviour.

Depending on our place within an institution's social hierarchy, we are exposed to and embody different kinds of social practices and skills. Just as we're encouraged to adopt certain kinds of behaviour as children by being corrected or scolded through overt and subtle indicators of social approval or disapproval, as we move into new levels of education, new jobs, or new communities we are continuously exposed to and encouraged to adopt new kinds of thinking, perceiving, and acting. Contingent on our place in an institution's hierarchy we will adopt different social practices and will have more or less influence on the evolution of the institution's social practices.

Concerning positions in institutional social hierarchies, a core aspect of the adoption of new practices is the symmetry (or asymmetry) of power that exists in different roles in interactions with others within institutions (De Jaegher, 2013b). We can see this in examining the effects asymmetric power relations have when one is in a subordinate position to another person as that person's practices may be more influential on us than we would recognise or like. This may be as innocent as adopting the gestures or speaking habits of a new friend or romantic partner that we want to impress or it could be harmful in the sense of adopting toxic masculinity or patriarchal gender norms from our elders (Hancock & Rubin, 2015). Thus, in order to understand how we adopt certain social practices and develop particular social skills we need to examine not only which institutions we are a part of, but also our place in the social hierarchy of these institutions.

In regard to social hierarchies, and as was stated above, social norms embedded in the social practices of institutions modulate individuals' intersubjectivity by constitutively shaping both their ways of thinking and acting in the world. While some of these embodied habits picked up from interactions with others, and the institutions they represent, can be good, others can be bad. These embodied habits can be bad in so far as they run counter to an individual's own interests, are detrimental to their wellbeing, or lead to social disapproval from other social institutions and their representatives (Maiese, 2018).

Recognising the harmful potential of social practices and interactions with institutions is essential for understanding the relationships the neurodivergent have with neurotypical institutions. As I present in the next section, the asymmetry of power that often exists in autistic individuals' interactions with neurotypical individuals within neurotypical institutions can often be detrimental to the autistic individuals. In this regard, the enactive framework is especially useful as its focus on intersubjectivity can not only reveal the deep influence social practices and institutions have on cognition and identity, but also provide a systemic framework for investigating how social practices and institutions can be harmful.

Autism and neurotypical institutions

In this section, utilising enactivism I show the constitutive influence neurotypical social practices and institutions have on the development of autistic individuals' social cognitive skills and their wellbeing. I will adopt a social model approach to the examination of autism as autism is defined and determined by neurotypical societies' ways of moving, communicating, and thinking, which autistic individuals may do differently. It is important to note that autistic individuals manifest characteristics of autism differently and some of these characteristics involve cognitive strengths in comparison to neurotypical individuals. As Chown and Beavan (2011) highlight, some of these cognitive strengths include good rote memory skills, ability to assimilate information quickly, long-term information memory, and high levels of concentration on specialised interests. Though some individuals are always going to be different from the neurotypical norm, these differences do not need to prevent autistic individuals, or other neurodivergent individuals, from participating in society.

I begin by showing how the double empathy hypothesis and monotropism theory can be integrated with the enactive framework. This integration is mutually beneficial as enactivism provides a general framework for explaining cognition with its unique focus on intersubjectivity, and the double empathy hypothesis and the monotropism theory provide insights for understanding autism previously missing from enactive accounts.[2] Milton, Heasman, and Sheppard (2018) explain the double empathy problem as neurotypical and autistic individuals having differences in their sociality that leads to frequent misunderstandings when interacting with each other. As both parties have difficulty in interacting fluidly with one another, it is a 'double problem' as the difficulty does not rest solely with the neurotypical individual or the autistic individual. Milton et al. (2018) claim that as interactions unfold, the initial differences in social saliency that cause the double empathy problem can quickly lead to critical misunderstandings that can potentially terminate the interaction. Milton (2012a) claims the experience of encountering this kind of difficulty is more severe for neurotypical individuals than autistic individuals, as it is an uncommon experience for neurotypical individuals.

The explanatory scope of the double empathy hypothesis aligns with enactivism's focus on intersubjectivity as it considers both the individual dispositions of agents in interactions, and the social context in which interactions take place (Milton et al., 2018). Additionally, as Milton et al. (2018) claim, there is a deep connection between the monotropism theory and the double empathy hypothesis. They sum up the core claim of monotropism as an essential difference between monotropic individuals, whose tendency is to localise attentional resources on a specific interest while excluding other potential inputs, and polytropic individuals whose attentional resources are capable of spreading to multiple inputs simultaneously (Murray et al. 2005, cited in Milton et al., 2018). Milton et al. (2018, p. 5) hypothesise that the kinds of reciprocal misunderstandings the double empathy problem highlights could be a 'consequence of a transactional, albeit socially situated, developmental process'. This is in line with enactivism's claim that embodied differences result in a different kind of intersubjectivity where the social world is experienced in structurally different ways. According to enactivism, this will in turn lead to the development of different kinds of social cognitive skills.

In regard to this, Murray et al. (2005, p. 140) claim that since social interactions and language development and use 'require broadly distributed attention', monotropic attention patterns produce a different kind of experience as a monotropic individual will be 'inhibited by the canalisation of available attention into a few highly aroused interests'. In line with enactivism's appeal to primary intersubjectivity as a set of basic capacities from which more sophisticated social skills develop, monotropic attention patterns would influence primary intersubjectivity in a structured way that can then be a basis for explaining the differences seen in autistic social cognitive development. This can be seen in monotropism's claim that learning new skills requires having an interest in doing so, and interests require both awareness and motivation, which are affected by monotropic attention patterns. Since monotropic attention patterns 'inhibit simultaneous awareness of different perspectives' this limits the intersubjective experience monotropic children have of being aware of others' viewpoints. This is especially true in regard to early language development.

Explaining monotropic differences in language development, Murray et al. (2005, p. 150) explain that conversations occur on multiple levels, such as 'phonetic (sound), phonological (rule-governed sound), syntactic (grammar), semantic (word and sentence meanings), and pragmatic (adjusted to each other's current interests)' through a sequence of events. For monotropic children, phonetic sounds may not be identified and connected to one another as they could be perceived as 'merely some among many noises in an unfiltered, undifferentiated aural environment' (p. 150). For a monotropic child, language needs to 'become an object of interest' otherwise the child may take longer to realise the meaningfulness of language (p. 150). Even after acquiring language skills, delays autistic children have in conversation (Leary & Donnellan, 2012) may violate the neurotypical norms, which often results in neurotypical individuals finding these long pauses uncomfortable and attempting to change the subject or drop the conversation. This creates a reciprocal feedback loop where the difference in attention, rhythm, and coordination makes social interaction difficult for both parties. However, these experiences are more detrimental for the autistic child as the child then loses opportunities to socially interact, which is important both for developing more sophisticated social cognitive skills and has a harmful psychological impact.

With respect to the importance enactivism places on synchronisation and coordination for social interactions influencing the development of culturally specific practices, the monotropism theory clarifies autistic differences. Murray et al. (2005, p. 144) claim that 'shortage of attention is key to the lack of simultaneous activity, rather than a lack of synchronization per se'. As synchronisation in rhythm and coordination were shown in the previous section to be important for maintaining fluidity in social interactions in order to pick up social practices, attention pattern differences that lead to difficulty in synchronisation would make picking up new social practices more difficult. Additionally, as the double empathy hypothesis highlights, this synchronisation issue cuts both ways as the social world is differently salient for autistic individuals and neurotypical individuals, which can again lead to more frequent breakdowns in interaction and limit autistic individuals' opportunities to socially interact.

In regard to autistic sensory differences, Milton (2012b) claims that autistic individuals' hyper/hyposensitivities to sounds, lights, smells, and touch can also be partly explained by monotropism. As a monotropic individual is either attending just to a particular sensory stimulation itself (hypersensitivity) or attending to a different aspect of the environment to the extent of not noticing another stimulation (hyposensitivity), this produces differences from the neurotypical norm. While hyper/hyposensitivities vary between autistic individuals across the spectrum, these differences may be able to be better explained on a case-by-case basis by examining the individual's monotropic tendencies. Importantly, these differences in sensitivities may also affect social interactions autistic individuals have as the sensory environment may make it harder for them to focus on the many different vocal and gestural aspects of the other's behaviour, potentially missing important normative aspects.

Monotropic attention patterns can further explain why autistic individuals have difficulty in accessing 'socially-salient information needed to fit into and become responsively regulated by the expressive norms governing' neurotypical institutions and practices (Krueger & Maiese, 2019, p. 24). For example, autistic individuals may find it difficult to detect subtle differences in expressive style required in situations where one needs to recognise the specific intention of a smile being either 'cold, sarcastic, confident, or wry' (p. 24). However, as the double empathy problem hypothesises, this is a bi-directional problem. Milton et al. (2018) cite research (Sheppard, Pillai, Wong, Ropar, & Mitchell, 2016; Edey et al., 2016) that suggests neurotypical individuals similarly have difficulty with identifying facial expressions of autistic individuals and making sense of autistic individuals' behaviour in interactive contexts.

While autistic individuals have difficulty in smoothly participating in neuro-typical practices, Krueger and Maiese (2019) point out that autistic individuals have their own practices that neurotypical individuals have difficulty with recog-nising, accepting, and even participating in. These autistic practices can include the observable behaviour of 'self-stimulation', which includes 'hand-flapping, finger-snapping, tapping objects, repetitive vocalisations, or rocking back and forth' (p. 27). Self-stimulation is known to help autistic individuals 'adapt to and negotiate changing environments' by organising sensations that help 'manage the physical, perceptual, and emotional demands of a given situation' (Leary & Donnellan, 2012, p. 51, cited in Krueger & Maiese, 2019, p. 27). On a mono-tropic reading, self-stimulative behaviours may help to shift attention away from overwhelmingly intense stimuli in order for the autistic individual to be able to then refocus his/her attention to other aspects of the environment. In this way, practices like these can be very helpful and comforting for autistic individuals, as it is a way for them to modulate their experience of the environment and the significance different sensations have for them via a controllable embodied behaviour.

While these practices can assist an autistic individual by modulating his/her attention in order to relieve issues related to hypersensitivity, the practices may seem off-putting for neurotypical individuals. Neurotypical individuals have a hard time accepting autistic self-stimulatory practices in the sense that neuro-typical individuals may not know how to respectfully and appropriately respond when autistic individuals engage in self-stimulatory practices in social interac-tions. Neurotypical individuals' inability, or unwillingness, to appropriately respond to self-stimulatory practices not only further strains the coordination and rhythm of the current interaction, but it may result in the neurotypical individuals having less interest in interacting with autistic individuals in the future. These detrimental effects may occur because the neurotypical individuals do not see the positive roles these autistic practices have for autistic individuals (Krueger & Maiese, 2019). Without a proper awareness of the significance the practices have for an autistic individual, neurotypical institutions and their representatives may view these practices negatively. This could in turn make interactions with the institution even more difficult for the autistic individual.

Nevertheless, this need not be the case. If neurotypical individuals are informed of the value self-stimulatory behaviours have for autistic individuals, it is possible for these detrimental effects to be avoided. In fact, neurotypical individuals can even adopt or engage in autistic practices. By doing so, it is possible to alter neurotypical institutions to create more space for autistic individuals. For example, some institutions, such as Manchester University's Student Union, have adopted hand flapping instead of clapping in order to thank speakers in an effort to be more inclusive for both deaf students and autistic students (Hinsliff, 2018). As this example shows, not only can awareness of autistic practices improve autistic individuals' interactions, but awareness can also help create space for not just acceptance, but even the adoption, of autistic practices by neurotypical individuals and institutions.

Though the aforementioned example shows the possibility of making neurotypical institutions more inclusive, the preceding intersubjective differences more often than not result in autistic individuals having more difficulty in smoothly participating in the everyday practices of neurotypical institutions. This can be partly explained by these differences making it more difficult for autistic individuals to detect the normative components of institutional practices (Krueger & Maiese, 2019). Conversely, in accordance with the double empathy hypothesis, this also means that it is more difficult for neurotypical individuals to detect the normative components of autistic social practices, such as self-stimulatory behaviours or hand flapping instead of clapping, and the importance these autistic practices have for autistic individuals. Difficulty in recognising these differences in each other's practices makes conforming to each other's expectations more difficult for both autistic individuals and neurotypical individuals.

The enactive framework's focus on intersubjectivity coupled with the contributions from the double empathy hypothesis and monotropism theory offers a systematic way of examining an autistic individual's experiences interacting with social practices and institutions. Understanding these intersubjective differences is important because even though autistic individuals jointly inhabit the same neurotypical institutions as neurotypical individuals, autistic individuals' difficulties in smoothly participating in the institutions' social practices can lead to a stigma that lowers their status in the institutions' social hierarchy. This means that in the context of institutions and their practices, autistic individuals end up feeling more isolated and alienated from not only the institution, but also from the people within the institution. Sarrett's (2018, p. 687) survey of autistic students in Australian universities found that 'only 27% reported having their social needs met.' Additionally, Gelbar, Shefcyk, and Reichow's (2015) literature review found that of the autistic students surveyed, '56% reported feeling lonely, 61% reported feeling isolated, and 42% reported feeling depressed' (cited in Sarrett, 2018, p. 687). These effects can be even more pronounced in institutions that have a social hierarchy that even more strongly prioritises the interests or needs of individuals who are regarded as better performers than their peers, such as is common in occupational institutions.

This leads back to issues raised in the last section, that there can be an asymmetry of power between interactors in interactions within social practices and institutions, and that some practices and institutions may lead to harmful habits that can be detrimental to one's wellbeing. For example, in a university educational setting there is already an asymmetry of power between a student and the professor that makes it difficult for many neurotypical students to speak up during in-class discussions, but for autistic students this asymmetry of power is even greater because of the difficulties discussed above. This can lead to autistic students feeling less confident with speaking to professors or speaking up during in-class discussion (Sarrett, 2018). However, when this occurs it only reinforces feelings of isolation and reduces autistic students' ability to practice the kind of social cognitive skills that in-class discussions are meant to help develop along with learning the course content. In line with the enactive framework, these kinds of experiences and interactions can have a negative impact by constitutively shaping autistic individuals' social cognitive habits to avoid these kinds of experiences and interactions.

According to enactivism, we develop our identities and ways of thinking and being through intersubjectivity, that is, our interactions with others, social institutions, and their social practices. Nevertheless, institutions 'cultivate framing patterns' and constitutively shape embodied cognitive habits even if these ways of thinking and being are counter to individuals' explicit interests or are in other ways harmful to their wellbeing (Maiese, 2018, p. 16). The above examples show how neurotypical institutions can have these detrimental effects on autistic individuals in particular, but also on the neurodivergent in general. Thus, through examining the intersubjective aspects of social hierarchies and asymmetrical power relations it is possible to further understand how certain social practices can be harmful. Addressing these aspects is necessary not only for improving the overall wellbeing of autistic individuals, but to also improve interactions between neurotypical individuals and autistic individuals. In order to improve the interactions between neurotypical individuals and autistic individuals in shared social institutions, there needs to be a shift in the focus of interventions away from autistic individuals towards the intersubjective realm of neurotypical social practices and institutions.

Conclusion

This chapter has shown how the enactive framework, utilising a social model approach, can provide a systematic method to develop comprehensive explanations of autistic individuals' intersubjective relationships with neurotypical social practices and institutions. If the analysis presented here of the ways in which neurotypical social practices and institutions can harm autistic individuals' social cognitive skills, identity, and wellbeing is on the right track, then we have good reasons to think that we should shift the focus of interventions away from the neurodivergent individual him/herself towards the social environment of neurotypical social practices and institutions.

One potential research path for developing interventions targeting the social environment that enactivism can assist with is the developing field of animal-assisted therapy for autistic individuals to help improve their social skills and wellbeing (see Smith, 2018). While understanding human institutions and social practices are crucial for understanding cognition, this is only one aspect of the social world most individuals engage with. An advantage of adopting the enactivist framework in relation to animal-assisted therapy is that enactivism has through its focus on intersubjectivity the capacity to explain both the nature of this kind of interspecies engagement and how it can be helpful for autistic individuals. For example, therapy trained dogs are not only capable of primary intersubjectivity, but also a basic form of secondary intersubjectivity. As enactivism embraces the diversity of cognition in such a way that is not bound to a particular species, enactivism is uniquely suited for examining and explaining the nature and impact of non-human animal interactions on neurotypical and neurodivergent individuals' intersubjective capacities and cognitive skills.

A second research path in need of further exploring is the development of an enactive neurodiversity paradigm for education. Such a paradigm would educate students about neurodiversity as a property of people in general, rather than singling out particular neurodivergent people for what many people consider to be 'special treatment'. An enactive neurodiversity paradigm approach in education would essentially involve giving a prominent place in the education system for understanding neurodivergent differences by educating students about intersubjectivity and the differences individuals have in their experiences of the world based on their embodied and embedded differences. This means teaching children explicitly about how to understand differences and see the value in having differences. For this reason, an enactive neurodiversity paradigm approach towards education should be extended to all levels of education, from primary education onwards, with the teaching developing in complexity from level to level in the usual manner. Through further developing these research paths for interventions, and by implementing the interventions, we can adjust the social world co-inhabited by neurodivergent individuals and neurotypical individuals in order to make it more inclusive for the neurodivergent and improve their overall wellbeing.

Notes

1 The term 'social practices' is used here denote the patterns of behaviour one adopts in social environments. In this sense, social practices can develop either through exposure

to cultural patterns of behaviour or from a pattern of behaviour an individual establishes over time in response to certain kinds of social stimuli.

2 While the monotropism theory may only be relevant to explanations of autism, the double empathy hypothesis can also explain the social difficulties experienced by non-autistic neurodivergent individuals. This is because other forms of neurodivergence will also affect an individual's relationships to neurotypical social institutions and interactions with neurotypical individuals. This is partly due to these neurodivergent individuals having differences in coordination capacities and having to face bias issues in interactions with neurotypical institutions and individuals.

References

Chown, N., & Beavan, N. (2011). Intellectually capable but socially excluded? A review of the literature and research on students with autism in further education. *Journal of Further and Higher Education, 36*(4), 477–493.

De Jaegher, H. (2013a). Embodiment and sense-making in autism. *Frontiers in Integrative Neuroscience, 7*(15), 1–19.

De Jaegher, H. (2013b). Rigid and fluid interactions with institutions. *Cognitive Systems Research, 25*(26), 19–25.

De Jaegher, H. (2018). The intersubjective turn. In Newen, A., De Bruin, L. & Gallagher, S. (Eds.) *The Oxford handbook of 4E cognition* (pp. 453–467). Oxford: Oxford University Press.

Edey, R., Cook, J., Brewer, R., Johnson, M. H., Bird, G., & Press, C. (2016). Interaction takes two: Typical adults exhibit mind-blindness towards those with autism spectrum disorder. *Journal of Abnormal Psychology, 125*(7), 879–885.

Fenton, A., & Krahn, T. (2007). Autism, neurodiversity, and equality beyond the 'normal'. *Journal of Ethics in Mental Health, 2*(2), 1–6.

Fuchs, T. (2015). Pathologies of intersubjectivity in autism and schizophrenia. *Journal of Consciousness Studies, 22*(1–2), 191–214.

Gelbar, N. W., Shefcyk, A., & Reichow, B. (2015). A comprehensive survey of current and former college students with autism spectrum disorders. *The Yale Journal of Biology and Medicine, 88*(1), 45–68.

Hancock, A. B., & Rubin, B. A. (2015). Influence of communication partner's gender on language. *Journal of Language and Social Psychology, 34*(1), 46–64.

Hinsliff, G. (2018). Jazz hands – now that's an act of kindness I can applaud. *Guardian*, 6 October. Available at: www.theguardian.com/commentisfree/2018/oct/06/jazz-hands-manchester-clapping (Accessed: 22 October 2019).

Jurgens, A., & Kirchhoff, M. D. (2019). Enactive social cognition: diachronic constitution & coupled anticipation. *Consciousness and Cognition, 70*, 1–10.

Krueger, J., & Maiese, M. (2019). Mental institutions, habits of mind, and an extended approach to autism. *Thaumàzein, 6*(0), 10–41.

Leary, M. R., & Donnellan, A. M. (2012) *Autism: Sensory-movement differences and diversity*. Cambridge, WI: Cambridge Book Review Press.

Maiese, M. (2018). Life shaping, habits of mind, and social institutions. *Natureza Humana-Revista Internacional de Filosofia e Psicanálise, 20*(1), 4–28.

Milton, D. (2012a). On the ontological status of autism: The 'double empathy problem'. *Disability & Society, 27*(6), 883–887.

Milton, D. (2012b) *So what exactly is autism?* London: Autism Education Trust.

Milton, D., Heasman, B., & Sheppard, E. (2018). Double empathy. In Volkmar, F. (ed.), *Encyclopedia of autism spectrum disorders* (pp. 1–7). New York: Springer.

Murray, D., Lesser, M., & Lawson, W. (2005). Attention, monotropism and the diagnostic criteria for autism. *Autism, 9*(2), 136–156.

Nagy, E., Pilling, K., Watt, R., Pal, A., & Orvos, H. (2017). Neonates' responses to repeated exposure to a still face. *PLoS One, 12*(8), 1–21.

Roepstorff, A., Niewöhner, J., & Beck, S. (2010). Enculturing brains through patterned practices. *Neural networks: The Official Journal of the International Neural Network Society, 23*(8–9), 1051–1059.

Sarrett, J. C. (2018). Autism and accommodations in higher education: insights from the autism community. *Journal of Autism and Developmental Disorders, 48*(3), 679–693.

Sheppard, E., Pillai, D., Wong, G. T. L., Ropar, D., & Mitchell, P. (2016). How easy is it to read the minds of people with autism spectrum disorder? *Journal of Autism and Developmental Disorders, 46*(4), 1247–1254.

Smith, T. (2018). Does animal assisted therapy (AAT) positively increase socialization skills in children diagnosed with Autism Spectrum Disorder (ASD) compared to no interaction with animals? *PCOM Physician Assistant Studies Student Scholarship, 312*, 1–14.

Part III

Cross-neurotype communication

Part III

Cross-neurotype
communication

Chapter 6

Neurodiversity and cross-cultural communication

Alyssa Hillary

Cross Cultural Communication/跨文化交际

Too blunt.	太直接了。
It'd be insulting.	他会收到侮辱。
They'd take it personally.	就是个人侮辱！
You can't just *say* that.	你不可以这样说（语法没问题，而不够委婉。）
You have to hint.	暗示一下（暗示不应该那么明显！）
Be subtle.	你可能需要说的委婉一些。

It was the same meaning –
Almost the same words.
It was the same bluntness –
Even the same confusion.

Then	But.
I claim a cultural difference.	They claim a cultural difference.
Autistic and Neuronormative.	American and Chinese
Denied.	Known issue.
Autism doesn't get a culture.	The reason for today's lesson.

Autistic people are too blunt.	Americans are blunt.
It's because we're disabled.	Chinese people are subtle.
We need to be 'fixed.'	It's a cultural difference.

美国人直接？ (Americans are blunt?)
是可笑的！ (That's laughable!)
自闭症者直接。 (Autistics are blunt.)
美国人委婉。 (Americans are subtle.)
中国人更委婉。 (Chinese people are even subtler.)

I have a communication disability.	There is a cultural difference.
This is my problem.	We can work together.

My teacher says it's different, never having listened.
She's never watched the Autistic version of this conversation.
Not that she'll admit.
(She's been the neuronormative side.)
(She thought she was only the Chinese side.)
(I knew she was both.)

Neurodiversity and cross-cultural communication

Both cultural and neurocognitive diversity exist, and communication across cultural and neurocognitive differences is part of life. In both cases, people must communicate across significant differences. For this reason, and because neurodivergent cultures exist, I will discuss connections between these forms of communication across difference.

To this end, both neurotypes and cultures require definition.

A neurotype describes a cluster of similar neurological and cognitive ways of being. Autism, for example, is a neurotype, as is aphantasia, the lack of voluntary imagery. The placement of divisions between neurotypes is a subject of debate (Leong, Hedley, & Uljarević, 2019), as the validity of clustering at all may also be (John, 2018). Neurodiversity movements typically consider neurological differences or differences in 'brain wiring' to describe something real about identity and neurocognitive experience, with both a diversity of brains and of minds and experiences. On the other hand, user/survivor discourse often considers brains in terms of reductionism and pathologisation (Jones & Kelley 2015). Self-identification as 'Mad' doesn't specify a neurocognitive cluster, but rather a commonality of pathologised experience. Coming from neurodiversity movements, I use 'neurotype' for all such groupings, including neurotypicals, but recognise this may not be ideal. I believe connections and overlaps between communication between people of different neurotypes and people of different cultural backgrounds will hold up, even if the current clustering (set of named neurotypes and related identities) or clustering itself does not. I also believe communities built around current identities can be discussed as cultures whether or not these identities match neuroscientifically coherent clusters.

Culture does not have a single definition (Apte, 1994; Peters, 2000), but I use a broad conception of culture, combining historical, linguistic, social, political, personal, and aesthetic aspects (Peters, 2000). This supports the discussion of national (e.g. Chinese, Irish, or USA-ian), ethnic (e.g. African American, Ashkenazi Jewish, or Uighur), and group (e.g. youth, Disabled, or Autistic) cultures. Disability cultures' inclusion of "personal narratives, oral history, art, and other forms of knowledge produced by and for disabled people' (Acevedo, 2018, p. 124) supports discussion of neurodivergent cultures on these same bases.

Cross-cultural (or intercultural) communication describes both communication between people of different cultural backgrounds and the field of inquiry to support this communication. Ideally, cross-cultural communication considers all participating cultures as valid and supports communication without requiring any communicator to pretend to come from another communicator's culture. Unfortunately, racism, classism, and other systems of oppression mean this doesn't always happen. Popular discussions of African American Vernacular English as 'incorrect' language are one example of racist and classist failures to respect cultural communication styles. The residential school system used against Indigenous people in the United States of America and Canada (MacDonald, 2007) went far beyond communication styles but included them, banning Indigenous languages. People who experience both neurobigotry and racism describe overlap and parallels in these experiences, which deserves further discussion, preferably by people who can speak to both experiences (Brown, Ashkenazy, & Onaiwu, 2017; ObeySumner, 2018; Prahlad, 2017). However, this discussion primarily draws on cross-cultural communication in its successes, rather than in the places where forms of oppression beyond (and interacting with) ableism introduce additional barriers.

To support cross-neurotype communication, we could explicitly learn about difference and how to understand and communicate with people unlike ourselves – in both directions, rather than only teaching marginalised people to fit in. However, ideas of cross-cultural communication and mutual understanding are not always (or even usually) applied to communication between people of different neurotypes, such as between Autistic and non-autistic people, or between Mad people and those who are not considered mad. Instead, miscommunications between neurotypical people and neurodivergent people are often blamed on neurodivergent people's perceived 'social deficits,' including in 'Theory of Mind.' This may relate people in primarily abled spaces failing to consider (or even be aware of) disability cultures grown from mutual recognition (Acevedo, 2018), including people in primarily neurotypical spaces failing to understand Autistic perceptions and cultures (Leong, 2016; Milton, 2012). Leong inverts Theory of Mind discourse to assert neurotypicals can't understand autistic mental processes. Milton describes this breakdown as the 'double empathy problem', in which attempts by non-autistic (often neurotypical) actors empathise inaccurately with autistic people and then invasively insist upon their inaccurate results. This problem is based in the difference rather than either partners disposition (Milton, 2012). While descriptions vary, Autistic scholars question 'theory of whose mind?' and note neurotypical people misunderstand autistic minds (Harvey, 2016; Leong, 2016; Milton, 2012).

Having participated in conversations about social expectations and communication in cross-neurotype and cross-macro-cultural contexts, I have experienced connections between the two (Hillary, 2018). Others have also noted these parallels (Attwood, 2006; Miyazaki & DeChicchis, 2013). I therefore consider explicitly Neurodivergent cultures, including Autistic (Davidson, 2008; Dekker 1999; Parsloe, 2015), Mad (Spandler, Anderson, & Sapey, 2015), emerging Aphant (Kendle, 2017), and broadly Neurodivergent/Neurodiverse (Clagg & Rocha, 2018;

The Big Anxiety, 2017) cultures, as well as neurotype-dependent differences in perception, which affect shared understandings of experiences (Semiotic Spectrumite, 2018).

Like broader disability cultures (Acevedo, 2018), Neurodivergent cultures are themselves diverse, with differences between the Mad and Neurodiversity movements as one example. There is overlap between the two and space for conversation between them (Dekker, 1999; Graby, 2015), but they have different histories and trajectories. Aphants, or people who do not experience voluntary visual imagery, may use the language of neurodiversity and generally draw on that wider movement, but again have a different (and shorter) collective history than the Mad and Neurodiversity movements – unlike Mad and neurodivergent people, Aphants get to collaborate on research about aphantasia, and their insights on aphantasia actually get taken seriously as things to consider. This is not generally true when scientists study many other forms of neurodivergence. Aphants have been recognised as research partners from near the neurotype's naming (Kendle, 2017).

Even within one neurotype, such as autism, variety exists. Some people are shaped more by other disability movements while in turn shaping neurodiversity and Autistic groups (Baggs, 2020). Others note connections with both broader disability rights movements and psychiatric survivors movements (Dekker, 1999). Yet others seek alignment with only a certain subset of autistic people they consider to be 'higher functioning' or 'more able', reproducing neuronormative exclusion, or exclusion-based privileging neurotypicality. This aspie supremacy has dangerous consequences (Baggs, 2010), and assumptions of a single Autistic culture erases Autistic cultures that welcome people who do not meet arbitrary standards of functioning.

Another dimension of cultural variation within neurotypes is our home cultures. People's home cultures affect their understanding of neurodivergence, including different understandings of concussions between rural cultures where more people experience them and urban cultures where they are rarer and more strongly avoided (Clagg & Rocha, 2018). Cultural understandings of autism similarly mediate diagnosis and pathologisation (Papadopoulos, 2016). Then, in addition to being understood as functional neuro-cultures (Leong, 2016), Neurodivergent cultures, like broader Disabled cultures, can be understood as subcultures, countercultures, or co-cultures (Peters, 2000; 李, 2011). Then, if a French Autistic person, an American Autistic person, and a Chinese Autistic person are present in one Autistic cultural space, cross-cultural communication occurs because they are respectively French, American, and Chinese. French Autistic bloggers directly reference these differences, describing events from 'communauté autiste anglophone', or English-speaking autistic communities (dcaius, 2018). While some (white) people claim autism as a nationality, race, or ethnicity, neurodivergent people already have these characteristics. Treating neurotypes as nationalities, races, or ethnicities erases national, racial, and ethnic diversity within and between Neurodivergent cultures and communities. It also hides the lack of such diversity from notice in homogeneous Neurodivergent communities – many descriptions of

Autistic culture describe a subset of white, western Autistic culture (Brown et al., 2017; ObeySumner, 2018; Onaiwu, 2020).

Additionally, while I apply principles of cross-cultural communication to practices I discuss as neuro-cultural practices and more direct neurocognitive effects, none are purely one or the other. Both culture and (neuro)biology affect all human knowledge, beliefs, and actions. One aspect may be more obvious at times, but both are always present. I seek to apply the mutual understanding of effective cross-cultural communication to cross-neurotype communication rather than continuing the pathologisation of neurodivergent communication styles, regardless of the balance of cultural and (neuro)biological influences.

I work from this basis, and from my own experiences studying abroad. Between 2009 and 2014, I studied abroad five times, for periods ranging from two weeks to an academic year. Most related to my participation in the International Engineering Program for Mandarin Chinese, designed to prepare engineers to practice cross-culturally in another language, the Chinese Flagship Program, designed to help students from any major reach fluency and cultural competence in Chinese, or both. During my last and longest experience in 天津，中国 (Tianjin, China), we had a unit on cross-cultural communication. We discussed both how my classmates and I, as Americans, could adapt to the ways Chinese people might be more 委婉 (subtle, approximately), as well as how Chinese people now expected (and tried not to be offended by) more 直接 (blunt, approximately) communication from Westerners. We focused on how we might learn to communicate better with Chinese people, since international structures of power mean Americans often expect to maintain social, cultural, and political dominance, and it was frankly our turn to meet Chinese culture somewhere closer to halfway. And yes, I think this parallels neuronormative expectations of dominance to the point that autistic people can't imagine what meeting us halfway would even mean (Asasumasu, 2011). At least our teachers expected us to remedy the imbalance in China.

We discussed a Western businessman who ordered bicycles from China. The bicycles made more noise than expected. The teachers asked how we might address the issue. We already knew that 'call the Chinese manufacturers and say the bikes are too loud' would be rude, as it was too blunt. I did not have additional suggestions, but my classmates did. One after the other, they suggested their ideas, and the teacher explained how they were still 太直接！(too direct!). Eventually, we gave up. The real answer? The Western businessman invited the head of the Chinese plant to lunch, and then to go cycling – on the noisy bicycles! The head of the Chinese plant heard the noisy bicycles for himself without being directly told and shortly fixed the problem. No one even directly said there had been a problem, but it was solved.

I would not have thought of this solution. I was subtler in China than I am in the USA – I was trying to be culturally appropriate. However, the difference between how subtle I could be and how subtle I was expected to be remained larger in China than in the USA. Despite the larger gap, I got far more understanding in China, where the cultural difference was expected, than I did in the

USA, where I'm simply expected to know better. This understanding included a greater acceptance of 'showing my work', in that I could say, 'I think X is the way to tell you I need Y, is that right?' and even if X wasn't quite right, I'd often get both an explanation of the correct script to use and the Y I actually needed. Even asking, 'I have no idea how to appropriately tell you I need Y, how should I say that?' worked, so it became my script when I wasn't sure there was a culturally typical way to express my needs. With understanding from all parties that a cultural difference was present, communication went much more smoothly, and this experience appears to be common among autistic people living abroad (assuming their macro-culture of origin gets this understanding in their new home country (Attwood, 2006).)

That is to say, the double empathy problem was less of a problem for me in China, despite the greater difference in perspectives, because principles of cross-cultural communication were used over the idea that 'I have a communication disorder, so this is my problem' (Asasumasu, 2011). Using principles of cross-cultural communication over neurotypically defined 'social deficits' challenges models that depend on these 'deficits' to explain communication difficulties between neurotypes.

The first such problematic model cites deficits in 'Theory of Mind', a term that conflates multiple skills, including an understanding that others could have different mental states and the ability to predict said others' mental states. The concept of Theory of Mind has been criticised in a variety of ways, including its complicated grammar (Harvey, 2016), its failure to account for autistic features in task creation and evaluation (Leong, 2016), and the unstated but logically deducible argument that autistic people are not actually human (Yergeau, 2013) and can't develop identities (Harvey, 2016) because of our supposed lack of Theory of Mind. That is, while discussions of Theory of Mind do not explicitly claim autistic people aren't human as Lovaas did (Chance, 1974), autistic inhumanity would logically result from an autistic lack of Theory of Mind and Theory of Mind as an essential human ability (Yergeau, 2013). Simon Baron-Cohen, a major figure in discussions of autism and Theory of Mind, makes both statements (Baron-Cohen, 1995). Similarly, discussions of identity formation and Theory of Mind lead to the logical conclusion that autistic people cannot form identities (Harvey, 2016). This contradicts the existence of Autistic cultural identities (Davidson, 2008; Dekker, 1999). Coming from the 'bias', if you'll call it one, that neurodivergent people are real human people with real human identities who do things for real human reasons, typical Theory of Mind discussions are a problem.

Daniel Hutto's Narrative Practice Hypothesis, which supposes people learn to understand the reasons for others' actions through stories that discuss the reasons for these actions (2003, 2012), challenges the concept of Theory of Mind. Applied to a neurodiversity paradigm understanding of neurodivergence and to neurodivergent cultures as (neuro-)cultures (Dekker, 1999; Leong, 2016) the hypothesis could support connections between cross-neurotype and cross-cultural communication. However, Hutto used a typical pathology view of autism, and so his application of the Narrative Practice Hypothesis to autism (2003) must be challenged by the use of cross-cultural communication principles in explaining and reducing difficulties in cross-neurotype communication.

Competing neuro-cultural practices

To discuss cross-cultural communication principles as applied to (neuro-)cultures and neurodivergence in general, some discussion of neurodivergent cultures and communities is valuable. Some such communities, such as Autistic communities and Aphant communities, rely heavily on the Internet (Dekker, 1999; Kendle, 2017; Parsloe, 2015). This is common for geographically scattered groups, and methods of studying online cultures have grown with the increasing recognition of online communities and cultures (Upward, McKemmish, & Reed, 2011). Some groups also meet offline, either having formed offline or temporarily gathering primarily online communities together in one space, where neuro-cultural practices may appear (Dekker, 1999).

Autistic neuro-cultural practices include both individual and collective stimming. I have both seen and participated in group expressions of flapping, jumping, and rocking, as well as back-and-forth repetitions of enjoyable sounds and words during my engagement with Autistic communities. Autistic autobiographical writing can be considered as a cultural practice due to the general consideration of personal narratives in disability cultures (Acevedo, 2018), the tendency for Autistic authors to connect their narratives to Autistic communities (Rose, 2005), and the cultural knowledge and expertise shared within these personal narratives (Hillary, in press). Other neuro-cultural practices include banning flash photography, accepting others' natural level of eye contact whether it is more or less than typically expected in the local normative culture, forming relationships that may not depend on or involve speech, greater comfort with silences (Dekker, 1999), both giving and listening to intense monologues about niche topics as a sign of interest and respect (Bertilsdotter Rosqvist, 2019), and adjusting the level and type of attention paid to nonverbal communications on an individual level to both include nonspeaking people and avoid misinterpreting people whose neurodivergent body languages we may struggle to interpret.

Neuronormative Western cultures, on the other hand, are more likely to demand eye contact as a sign of respect or attention, demand 'quiet hands' as a sign of being ready to learn (Bascom, 2011) and prioritise speech over other forms of communication. For neurotypical people to successfully communicate with neurodivergent people, they may need to reconsider their assumptions about what 'paying attention' and 'respect' look like, consider the utility of multimodal communication, and vary the level of attention and importance they give to nonverbal cues depending on their skill at reading potentially idiosyncratic cues. Autistic people seeking to communicate with neurotypicals, may find themselves at an advantage if participating in diverse Autistic spaces already taught them to accept the extent to which other people move or remain still, look at or away from communication partners, and use differing methods of communication (such as an increased reliance on speech.)

Some autistic communication may also involve generous assumptions about (or generous room for) common ground, with minimal concern when the suggested common ground is not, in fact, common and a reference is missed (Heasman & Gillespie, 2018). This allows space to find common ground unconnected, or only loosely connected, with previously known areas of mutual interest. In contrast, Western neurotypical communication may rely on narrower assumptions of common ground, with greater distress at the realisation that proposed common ground is not present. This distress may be part of the reason neuronormative communication relies so heavily on expectations of 'audience awareness', in which writers are expected to understand their audiences *potentially variable* mental states. Teachers may even suppose students are not, in fact, part of this neurotypical culture when they do not make use of this construct, reinforcing the neurotypical construction of and fascination with Theory of Mind (Yergeau, 2013).

Patient, consumer, survivor, and Mad movements, while superficially like both broader disability and neurodiversity movements, have their own community and cultural practices. These practices may include non-pathologising views of hearing voices and practices related to those voices (Mc-Carthy-Jones et al., 2015; Schrader, Jones, & Shattell, 2013), rejection of diagnostic labels in rebellion against the psychiatric system, the creation of Mad-positive music (Castrodale, 2019), and more recently, identification with disability to utilise treaties like the Convention on the Rights of People with Disabilities (Spandler, Anderson, & Sapey, 2015). Identification with group terms such as patient, consumer, survivor, and mad, rather than a specific diagnosis, both challenges diagnostic labels applied by professionals and recognises common experiences between people with similar experiences.

In addition to these explicitly cultural practices, neurotype-dependent differences in perception affect the creation of shared understandings of experiences (Semiotic Spectrumite, 2018). While neurotypical people from similar macrocultures may sometimes predict each other's thoughts or feelings by putting themselves in each other's shoes, people with significantly differing internal experiences or backgrounds can't do this accurately (Hutto, 2012; Semiotic Spectrumite, 2018). Learning to understand the actions of *different* others and their reasons through stories is somewhat possible. Doing so can support an understanding of why someone's actions make sense to them, if not predict their reasoning, but it does require a greater leap of understanding.

Prosopagnosia, or face-blindness, creates one example. Most people recognise others' faces. In fact, the 'famous faces test' is sometimes used to screen for dementia because most people recognise faces, and their errors on this test are related more to losing information about others than to facial processing (Hodges, Salmon, & Butters, 1993). For this reason, a neurotypical observer might assume that a person who doesn't recognise faces either does not care (lazy neurotypical) or is showing signs of dementia (a more commonly diagnosed, acquired/progressive form of neurodivergence). However, the ability to use faces to recognise people isn't shared across all neurotypes. To the extent that I recognise people, I use context, voice qualities, non-facial visual cues such as hair or clothing, and movement/postural patterns. Movement and postural patterns let me recognise some familiar people at a much greater distance than typical face-recognition would allow, but at the same time, I can fail to recognise a friend at a short distance after he gets a haircut or my own major professor because a symposium is not the context where I usually see her. Because the experience of how we recognise people is not shared between prosopagnosiacs and non-prosopagnosiacs, misunderstandings result. Someone who recognises faces typically may not understand why a face-blind person provides or requests information about clothing prior to a planned meeting, while someone who has learned to recognise people by their movement patterns may struggle to understand why a friend who depends on faces to recognise people requires closer proximity to recognise someone – unless the differences are explained.

Some Aphants, or people who don't experience voluntary visual imagery (no mind's eye/visual imagination) describe communication between people who visualise and people who do not in similar terms to cross-cultural communication. Visualisers may expect that tasks for which they use visualisation, such as mental rotation tasks (Vanderberg et al., 1978; Zeman et al., 2010), knowing what shape is formed by the hole in a capital A, knowing what colour eyes someone has, and drawing (Grinnell, 2016), therefore *require* visualisation. However, Aphants use alternative strategies, such as matching blocks and angles perceptually in mental rotation tasks (Zeman et al., 2010) or imagining the movement involved in drawing letters to understand their shapes (Grinnell, 2016). As in cross-cultural communication, these explanations take place across a significant difference, and a better understanding of how others experience the world can improve communication. People who know I can't visualise are less likely to propose guided visualisations for me, knowing these activities only create confusion or frustration. And I, in turn, can be reminded to make pictures and figures for my presentations or publications: *I* certainly don't rely on them, for all I create them in my neuroscience research and remind geometry students to draw them!

Finally, (neuro-)cultural practices and the direct effects of neurocognitive differences may be intertwined. One discussion of concussion as neurodiversity argues for the recognition of varying cognitive abilities related to both injury and opportunity as neurodivergent *and* that rural cultures where this variation is expected are therefore neurodiverse (Clagg & Rocha, 2018). In this case, cultural practices that increase the likelihood of concussion-related neurodivergent experiences, cultural practices that treat these experiences as a normal, expected part of life, *and* the direct effects of concussions on neurocognitive experiences interact. Urban communities assume we should avoid concussion risk. Rural communities may just accept that concussions can happen from the way they live. This is a culturally mediated difference and communication between these communities has to deal with the difference. (Clagg & Rocha, 2018). Similarly, communication between people with urban neuronormative expectations and rural people whose cognition has been affected by concussions may encounter difficulties related to neurocognitive differences (Clagg & Rocha, 2018).

In the absence of neurobigotry, or bigotry against neurodivergent ways of being, Hutto's Narrative Practice Hypothesis (Hutto, 2003) explains difficulties in understanding between neuro-cultural groups and provides a roadmap to improving communication. The hypothesis says that to understand why others act as they do, we must interact with stories that provide reasons for action (Hutto, 2003). Without such stories, difficulties are expected, and they may be partially remedied by interaction such stories. He claims,

> In listening to another's account we sometimes expand the scope of what we deem acceptable. This is normally achieved when the other fleshes out on a larger canvas why they took an action, sometimes introducing a different set of values, such that we are brought to see it in a new light. Or, more conservatively, we may at least begin to see why the action might 'make sense' to *them*, even if it still leaves us cold.
>
> (Hutto 2003, p. 347, emphasis in original)

There are, of course, caveats. Hutto himself notes that if a person's stated desires or stated reasons remain puzzling to a listener, then their actions will also remain puzzling, using the example of a desire to consume acorns (Hutto, 2003, 2012). While I have no desire to eat raw acorns, some Native Americans traditionally eat acorns, and preparation instructions for both acorn polenta and acorn flour exist online. Treating acorns as food isn't puzzling, though cultural context is useful. Hutto notes this lack of understanding of differing values or goals can cause confusion in communication between people of different backgrounds. If you read a story about a person who acts for reasons, it's easier to understand that story if you (1) might have similar reasons for action, and (2) would get similar effects from similar actions. Both conditions can be violated in cross-cultural communication and in cross-neurotype communication. These two conditions interfering with the understanding of narratives can explain a significant portion of so-called autistic difficulties understanding (neurotypical) others and understanding sophisticated (neurotypical) intentional actions and neurotypical difficulties in understanding neurodivergent people and our intentional actions (Leong, 2016; Milton, 2012). They also help explain difficulties in understanding between neurodivergent people of different neurotypes, who are rarely addressed in discussions of cross-neurotype communication.

Differences in priorities are at issue in conflicts between more individualistic cultures and more group-oriented cultures. People who will make personal sacrifices for the good of others may be confused by people whose priorities dictate that they take care of themselves first, and vice versa. Differences in goals are similarly a concern when a person who needs or prefers to avoid strobes for reasons related to their neurotype (common among people with sensory processing differences and epilepsy) finds their attempts to avoid strobes puzzle people who may *enjoy* strobe effects at concerts or clubs. It's mutual, of course, but ableism means the concerns of neurodivergent people are not given the same weight as neuronormative preferences. This can be true even when neurodivergent concerns and neuronormative preferences align, if we admit to our reasoning or others have presumed our neurodivergence.

Similarly, a story about looking at the same thing as another person to establish joint attention isn't helpful if we simply do not see the point in 'joint attention' – this neurotypical obsession with ensuring people are looking at the same object at the same time before telling each other about the object. The second consideration, the question of getting similar effects from similar actions, is also in effect when considering joint attention. Looking at someone to gauge where they are looking will not make sense if we can't tell where other people are looking! A student trying to look at the same thing as the teacher similarly won't lead to what neurotypicals expect of joint attention if the student can't tell where the teacher is looking.

Take, for example, social stories that claim we look at teachers to make them believe we are paying attention. For an autistic person who cannot simultaneously pay attention to what is being said and make eye contact, this reason for eye contact is patently absurd. Understanding this story requires a much larger leap to the perspective of someone significantly unlike ourselves than a story in which someone looks away from the teacher to pay better attention and avoid overloading the teacher!

The consideration of how an action's results change with the context also applies to cross-cultural communication. Making eye contact with authority figures is an expected display of respect in white Western cultures. It is *not* an expected display of respect in all cultures and doesn't necessarily have the effect of displaying respect to an authority figure. Telling a Chinese, Japanese, or Mongolian student to make eye contact with a teacher to show respect could easily be confusing: that's not the result they expect from eye contact, unless they are explicitly aware of the cultural difference. Even then, enforcing one culture's expectation over another may not be appropriate.

Or, consider the false-belief task, commonly used to test so-called Theory of Mind. Typically, a child sees a person put a toy in one location, either watches someone else move the toy to another location or is instructed to do so themselves, and is finally asked where the first person will think the toy is. When testing Samoan children, Mayer and Träuble ask children to move a toy to play a trick on a second child (2013). Stating this is a trick suggests the second child will *not* be told the toy was moved. Autistic criticisms of the test include the consideration that we don't *know* whether the second child (or adult, in some versions of the test) is in on the procedure (Blackburn, Gottschewski, & George, 2019). If the second person is an experimenter, it's not unreasonable to think they'd know the procedure! The criticism is considered valid enough to modify the experimental task when Mayer and Träuble work with Samoan children. It is *not* considered a true challenge to task validity when Autistic people 'fail' for this (or any) reason.

Another requirement in the use of stories in learning about how people act for reasons, of course, is that accounts must actually include reasons for acting, whether or not they are reasons a reader or listener would share. Samoans may assert that we cannot know what another thinks or feels (Robbins & Rumsey, 2008), and lo and behold: neurotypical Samoan children succeed in false belief tasks later than neurotypical American children (Mayer & Träuble, 2013). In this case, however, this is attributed to cultural differences in discussions of mental states, rather than an inherent inability to understand perspectives (Mayer & Träuble, 2013), and the authors do not follow autism researchers in arguing that later passing this test still, somehow, does not constitute a Theory of Mind (Baron-Cohen, 1995). In fairness, the Samoan children are exposed to fewer stories about people who act for reasons, which may better parallel exposure to fewer stories with neurodivergent reasons for acting (Bartmess, 2015) than the merely *nonsensical* stories Western neurodivergent people are exposed to, in which neurotypical actions and stated reasons fail to match actions with their effects on neurodivergent readers.

Concluding reflections

I therefore argue that both neurotypical difficulties in understanding neurodivergent reasons for acting, including autistic reasons for acting, and neurodivergent difficulties in understanding across neurotypes, including the reasons of *differently* neurodivergent people, depend on several factors. First, the differences in goals and effects of actions previously discussed apply in all directions. A neurotypical adult who fidgeted when they were bored will not immediately understand the actions of a neurodivergent person who fidgets because sitting still requires conscious attention. They can learn to understand this reason for acting, but requires a greater leap in perspective than simply assuming the neurodivergent person is also bored. Second, many stories about neurodivergent people are behaviourising rather than humanising. Bartmess discusses this tendency in terms of autism, noting that we don't act autistic 'because of "autism", full stop.' (Bartmess, 2015). However, behaviourising narratives write autistic characters that way. If neurodivergent characters are not presented as having full agency, and notably neurodivergent actions are presented as being because of neurodivergence, full stop, then these narratives *do not* provide reasons for acting. At this point, Hutto's Narrative Practice Hypothesis would *expect* people to fail to understand neurodivergent reasons for acting because they are not exposed to narratives in which neurodivergent people act for reasons.

However, rather than applying the same considerations to autistic social interaction that he mentions as possibilities in cross-cultural interaction, Hutto suggests autistic people cannot recognise that other perspectives exist. (I recognise that this perspective exists. I just think it's *terrible* and hope it's changed since 2012.[1]) This proposition engages in an unstated but deducible denial of autistic humanity (and likely neurodivergent humanity in general, were other forms of neurodivergence addressed) much as Theory of Mind discourse does (Yergeau, 2013). It additionally depends upon questionable statements about autism, such an autistic inability to engage in or comprehend pretend play (Hutto, 2003). As autistic children comprehend and engage in prompted pretend play (Jarrold, Boucher, & Smith, 1994), we don't actually lack this capacity (Harvey, 2016). Similar poor assertions, backed by research heavily criticised by autistic people for its neuronormative assumptions, are used repeatedly to uphold Hutto's proposition. In addition to these concerns, the proposition that autistic people, specifically, cannot recognise that other perspectives exist fails to address the mutuality of misunderstandings and social difficulties between autistic people and neurotypicals, as described by the double empathy problem (Milton, 2012).

In addition to these issues with the proposition that autistic people cannot understand the existence of other perspectives, the proposition is simply unnecessary. The Narrative Practice Hypothesis can cover difficulties encountered in cross-neurotype communication by considering social realities and similarities to cross cultural communication. Namely,

1. People of different neurotypes may have different goals and motivations.
2. People of different neurotypes may experience the results of the same action differently.
3. The standard stories told to children by which they learn about actions, goals, and motivations are currently *neurotypical* stories, often *white Western neurotypical* stories.
4. Stories about neurodivergent people frequently do not explain our visibly neurodivergent actions as involving goals, motivations, or reasons.

These considerations also appear in cross-cultural contexts. The study of both neurodivergent and neurotypical socialities may be better understood, and cross-neurotype communication may be improved, through learning from cross-cultural communication rather than studying neurotypically defined 'social deficits'. Sociality involves multiple people, so social difficulties cannot live within a single person.

Social stories directed at neurodivergent people exemplify the third consideration, rather than answering it – while these stories are directed at neurodivergent people, they work from neurotypical assumptions about the meanings and results of actions. However, the third and fourth considerations can be addressed through better neurodivergent narratives and *exposure* to said narratives. That is, representation matters. My discussion of Autistic auto/biographical writing notes the value of Autistic people reading for representation (Hillary, 2020), and these arguments apply generally for neurodivergent people. These same self-authored narratives with our reasons for acting also support outsider understandings of neurodivergent agency. Humanising neurodivergent fiction can similarly support understanding of neurodivergent people as active agents who act for reasons (Bartmess, 2015), as can Mad and Neurodivergent music (Castrodale, 2019). Both making more varied narratives available and improving access to existing narratives, particularly at intersections of multiple marginalised identities, can therefore support both cross-cultural and cross-neurotype communication.

In conclusion, Neurodivergent (sub)cultures exist, and treating them as (sub) cultures in terms of cross-cultural communication can improve cross-neurotype communication. This still applies to practices that are arguably less mediated by culture and more mediated by neurology. To the extent to which understanding the reasons for people's actions is desirable and possible, another question mediated by culture, a non-ableist application of Hutto's Narrative Practice Hypothesis, suggests that narratives that explain neurodivergent people's reasons for acting can support understanding of neurodivergent actions across neurotypes.

Note

1 Editors' note: Hutto confirms that the text in his 2008 book, which amounts to a blanket statement that the idea of divergent cognitive perspectives eludes autistic people, was never a good representation of his actual views, not even as they stood back in 2008. Hutto would now frame the passages in question quite differently than he did in 2008 in order to better capture his current, more nuanced views on, inter alia, interpretation of false belief test results, perspective taking, and propositional attributions (Hutto, 2020). The full text of the author/editor personal communication is available via email from Nick Chown.

References

Acevedo, S. M. (2018). *Enabling geographies: Neurodivergence, self-authorship, and the politics of social space* (Unpublished doctoral dissertation). California Institute of Integral Studies.

Asasumasu, K. (2011). What would meeting you halfway be?" *Radical Neurodivergence Speaking.* 11 April. http://timetolisten.blogspot.com/2011/04/what-would-meeting-you-halfway-be.html.

Apte, M. (1994). Language in sociocultural context. In R. E. Asher (Ed.), *The Encyclopedia of language and linguistics Vol. 4, 2000–2010.* Oxford: Pergamon Press.

Attwood, T. (2006). *The complete guide to Asperger's syndrome.* London: Jessica Kingsley.

Baggs, M. (2010, March 7). Aspie supremacy can kill. *Ballastexistenz.* URL: (Retrieved December 2019) https://ballastexistenz.wordpress.com/2010/03/07/aspie-supremacy-can-kill/

Baggs, M. (2020). Losing. In S. K. Kapp (Ed.), *Autistic community and the neurodiversity movement: Stories from the frontline* (pp. 77–86). London: Palgrave Macmillan.

Baron-Cohen, S. (1995). *Mindblindness: An essay on autism and theory of mind.* Cambridge, MA: MIT Press.

Bartmess, E. (2015, April 14). Writing autistic characters: Behaviorizing vs. Humanizing approaches. *Disability in kidlit.* URL: (Retrieved December 2019) http://disabilityinkidlit.com/2015/04/14/writing-autistic-characters-behaviorizing-vs-humanizing-approaches/

Bascom, J. (2011, 5 October). Quiet hands. *Just stimming*. URL: (Retrieved December 2019) https://juststimming.wordpress.com/2011/10/05/quiet-hands/

Bertilsdotter Rosqvist, H. (2019). Doing things together: Exploring meanings of different forms of sociality among autistic people in an autistic work space. *Alter; European Journal of Disability Research; Journal Europeen de Recherche Sur le Handicap*, 13(3), 168–178.

Brown, L. X., Ashkenazy, E., & Onaiwu, M. G. (Eds.). (2017). *All the weight of our dreams: On living racialized autism*. Lincoln, NE: DragonBee Press.

Blackburn, J., Gottschewski, K., & George, E. (2019). A discussion about Theory of Mind: From an autistic perspective from Autism Europe's Congress 2000. *Autonomy, the Critical Journal of Interdisciplinary Autism Studies*, 6(1).

Castrodale, M. A. (2019). Mad studies and mad-positive music. *New Horizons in Adult Education and Human Resource Development*, 31(1), 40–58.

Chance, P. (1974). A conversation with Ivar Lovaas. *Psychology Today*, 7(8), 76–80.

Clagg, D., & Rocha, S. D. (2018). Contrasting political ontologies of neurodiversity in high-concussion-risk rural cultures. *Journal of Curriculum Theorizing*, 33(1), 1–21.

Davidson, J. (2008). Autistic culture online: virtual communication and cultural expression on the spectrum. *Social & Cultural Geography*, 9(7), 791–806.

dcaius. (2018, July 24). Qu'est-ce que le « masking » autistique? *Survivre & s'épanouir: le blog de dcaius*. URL: (Retrieved December 2019) https://dcaius.fr/blog/2018/07/masking/

Dekker, M. (1999, November). On our own terms: Emerging autistic culture. In *Conferencia en línea*. www.autscape.org/2015/programme/handouts/Autistic-Culture-07-Oct-1999.pdf (Retrieved April 10, 2020).

Graby, S. (2015). Neurodiversity: Bridging the gap between the disabled people's movement and the mental health system survivors' movement. In H. Spandler, J. Anderson, & B. Sapey (Eds.) *Madness, distress and the politics of disablement* (pp. 231–244). Bristol: Policy Press.

Grinnell, D. (2016). Blind in the mind. *New Scientist*, 230(3070), 34–37.

Harvey, S. T. (2016). *A rhetorical journey into advocacy* (Unpublished master's thesis). St. Cloud State University.

Heasman, B., & Gillespie, A. (2018). Perspective-taking is two-sided: Misunderstandings between people with Asperger's syndrome and their family members. *Autism*, 22(6), 740–750.

Hillary, A. (2018). Who is allowed? In N. Walker & A. M. Reichart (Eds.), *Spoon knife 3: Incursions* (pp. 173–186). Fort Worth: TX: Autonomous Press.

Hillary, A. (2020). Autist/Biography. In Parsons, J. M. & A. Chappell (Eds.) *The Palgrave handbook of auto/biography* (pp. 315–339). https://doi.org/10.1007/978-3-030-31974-8.

Hodges, J. R., Salmon, D. P., & Butters, N. (1993). Recognition and naming of famous faces in Alzheimer's disease: A cognitive analysis. *Neuropsychologia*, 31(8), 775–788.

Hutto, D. D. (2003). Folk psychological narratives and the case of autism. *Philosophical Papers*, 32(3), 345–361.

Hutto, D. D. (2012). *Folk psychological narratives: The sociocultural basis of understanding reasons*. Cambridge, MA: MIT Press.

Hutto, D. D. (2020). Personal communication with one of the editors (Nick Chown).

Jarrold, C., Boucher, J., & Smith, P. K. (1994). Executive function deficits and the pretend play of children with autism: A research note. *Journal of Child Psychology and Psychiatry*, 35(8), 1473–1482.

John, Y. (2018). Answer to: Can you list all the different neurotypes? *Quora*. URL: (Retrieved December 2019) www.quora.com/Can-you-list-all-the-different-neurotypes-I-m-struggling-to-find-them-but-I-know-of-the-neurotypical-brain-psychopathic-brain-autistic-brain-dyslexic-brain-and-ADHD-brain-What-other-brains-are-there/answer/Yohan-John

Jones, N., & Kelly, T. (2015). Inconvenient complications: On the heterogeneities of madness and their relationship to disability. In H. Spandler, J. Anderson, & B., Sapey (Eds.), *Madness, distress and the politics of disablement* (pp. 43–56). Bristol: Policy Press.

Kendle, A. (2017). *Aphantasia: Experiences, perceptions, and insights*. Oakamoor, UK: Dark River Press.

Leong, D. S. M. (2016). *Scheherazade's sea – autism, parallel embodiment and elemental empathy* (Unpublished doctoral dissertation). University of New South Wales, Sydney.

Leong, D., Hedley, D., & Uljarević, M. (2019, e-pub). Poh-tay-toe, Poh-tah-toe: Autism Diagnosis and Conceptualization. *Journal of Child Neurology*. doi: 10.1177/088307 3819887587.

李志远. (2011). *中国残疾人共文化群体与主流非残疾人文化群体的跨文化非语言交际* (Unpublished master's thesis). 哈尔滨工程大学)

McCarthy-Jones, S., Castro Romero, M., McCarthy-Jones, R., Dillon, J., Cooper-Rompato, C., Kieran, K., ... & Blackman, L. (2015). Hearing the unheard: An interdisciplinary, mixed methodology study of women's experiences of hearing voices (auditory verbal hallucinations). *Frontiers in psychiatry*, *6*, 181.

MacDonald, D. (2007). First Nations, residential schools, and the Americanization of the Holocaust: Rewriting Indigenous history in the United States and Canada. *Canadian Journal of Political Science/Revue canadienne de science politique*, *40*(4), 995–1015.

Mayer, A., & Träuble, B. E. (2013). Synchrony in the onset of mental state understanding across cultures? A study among children in Samoa. *International Journal of Behavioral Development*, *37*(1), 21–28.

Milton, D. E. (2012). On the ontological status of autism: the double empathy problem. *Disability & Society*, *27*(6), 883–887.

Miyazaki, Y., & DeChicchis, J. (2013). Through the glass ceiling: A comparison of autistics and foreigners in Japan. *総合政策研究*, *42*, 31–40.

ObeySumner, C. (2018, 5 December). Black autistics exist: An argument for intersectional disability justice. *South Seattle Emerald*. URL: (Retrieved December 2019) https://southseattleemerald.com/2018/12/05/intersectionality-what-it-means-to-be-autistic-femme-and-black/

Onaiwu, M. G. (2020). 'A dream deferred' no longer: Backstory of the first autism and race anthology. In S. K. Kapp (Ed.), *Autistic community and the neurodiversity movement: Stories from the frontline* (pp. 243–252). London: Palgrave Macmillan.

Papadopoulos, C., 2016. Autism stigma and the role of ethnicity and culture. *Network Autism*.

Parsloe, S. M. (2015). Discourses of disability, narratives of community: Reclaiming an autistic identity online. *Journal of Applied Communication Research*, *43*(3), 336–356.

Peters, S. (2000). Is there a disability culture? A syncretisation of three possible world views. *Disability & Society*, *15*(4), 583–601.

Prahlad, A. (2017). *The secret life of a black aspie: A memoir*. Fairbanks, AK: University of Alaska Press.

Robbins, J., & Rumsey, A. (2008). Introduction: cultural and linguistic anthropology and the opacity of other minds. *Anthropological Quarterly*, *81*, 407–420.

Rose, I. (2005). Autistic autobiography: Introducing the field. In *Proceedings of the Autism and Representation: Writing, Cognition, Disability Conference*. URL: (Retrieved December 2019) https://case.edu/affil/sce/Representing%20Autism.html

Schrader, S., Jones, N., & Shattell, M. (2013). Mad pride: Reflections on sociopolitical identity and mental diversity in the context of culturally competent psychiatric care. *Issues in Mental Health Nursing, 34*(1), 62–64.

Semiotic Spectrumite. (2018, January 26). The belief in a theory of mind is a disability. URL: (Retrieved December 2019) https://semioticspectrumite.wordpress.com/2018/01/26/the-belief-in-theory-of-mind-is-a-disability

Spandler, H., Anderson, J., & Sapey, B. (Eds.), *Madness, distress and the politics of disablement* (pp. 43–56). Bristol: Policy Press.

The Big Anxiety. (2017). Neurodiverse-city. URL: (Retrieved December 2019) www.thebiganxiety.org/event-category/neurodiverse-city/

Upward, F., McKemmish, S., & Reed, B. (2011). Archivists and changing social and information spaces: a continuum approach to recordkeeping and archiving in online cultures. *Archivaria, 72*, 197–237.

Vandenberg, S. G., & Kuse, A. R. (1978). Mental rotations, a group test of three-dimensional spatial visualization. *Perceptual and Motor Skills, 47*(2), 599–604.

Yergeau, M. (2013). Clinically significant disturbance: On theorists who theorize theory of mind. *Disability Studies Quarterly, 33*(4).

Zeman, A. Z. J, Sala, S. D., Torrens, L. A., Gountouna, V.-E., McGonigle, D. J., & Logie, R. H. (2010). Loss of imagery phenomenology with intact visuo-spatial task performance: A case of 'blind imagination'. *Neuropsychologia, 48*(1), 145–155.

Chapter 7

Understanding empathy through a study of autistic life writing

On the importance of neurodivergent morality

Anna Stenning

Introduction

The notion that autism is defined by empathy deficits (and the related ideas of an absent Theory of Mind (ToM), otherwise known as mindblindness), has been used to suggest that autistic people are not fully moral (Barnbaum, 2008). As scholars and activists have observed in connection to cognitive theories about autism in general, autistic people have been denied characteristics that are commonly considered part of what it is to be fully human, including empathy, morality, a sense of self, imagination, narrative identity, integrity; introspection, self-hood, personhood; rhetoricity, gender, meaning-making, sociality, or flourishing (McDonagh, 2013; Milton, 2012; Rodas, 2018; Yergeau, 2018). They show how, in each case, these limitations are based on foreshortened or even non-standard definitions of these qualities, to ensure that they only apply to a cultural 'in-group'. This impoverishes the generalisability of any empirical or theoretical research that relies on it. These assertions become harder to sustain as more prominent autistics (e.g. Temple Grandin, Chris Packham, Greta Thunberg, Hannah Gadsby) enter the public arena and make valuable contributions to discussions about the nature of an ethical human life, and to what it means to be neurodivergent.

Within the academic realm, the ethical implications of human neurodivergence are far from well understood, and yet it is on this basis that funding and interventions are decided. While this may seem purely a 'theoretical' exercise within an academic essay, I believe that granting ethical value to neurodivergent people must happen both top down (challenging established theory and methodology) and bottom up (from experience), to have a chance to impact on society more widely. It is hoped that this chapter will be of some practical help to scholars who genuinely understand the value of including neurodivergent voices in both the methodological and ethical justifications for their work. While this kind of inclusion is often tokenistic and based on a shallow understanding of co-production or impact, much 'ethical work' needs to be done to question why it is happening in such ways. Within the field of autism research, I offer the following initial exploration.

Simon Baron-Cohen is the theorist most responsible for the association of autism and empathy deficits in the popular imagination. His idea of empathy is a propensity to 'naturally and spontaneously [tune] into someone else's thoughts and feelings, whatever these might be' (2003, p. 21). He believes that this is absent or impaired in autistic people. On the other hand, the literary critic Patrick McDonagh – as part of the first wave of critical autism studies within the humanities and social sciences that was willing to grant autistic voices some authority – observed that 'many autistic people assert that they do experience empathy' and this includes overwhelming empathy for other people and other species (2013, pp. 155–156). McDonagh considers that empathy, in Baron-Cohen's 'cognitive' sense, has been taken as a necessary basis upon which economic and social transactions take place. However, he notes that despite being depicted as a quality that is essential to humanity, empathy has no single characterisation through history. He concludes, therefore, 'empathy is an abstraction, a reification; any definition is bound to be the sum of a cluster of responses that someone (or some culture) defines *a priori* as "empathic"' (p. 47). Indeed, as we will discuss in detail, the question of what empathy is even within autism is significantly more nuanced and complex than Baron-Cohen's characterisation suggests. And it is interesting to note that, from McDonagh's writing to the present, humanities scholarship has retained an interest in autistic empathy in connection to our supposed affinity with other species (see, e.g. Figueroa, 2017).

And yet within the humanities, the prevalence of deficits-based models of autism is perhaps most problematically demonstrated by Deborah R. Barnbaum's *The Ethics of Autism: Among Them, But Not of Them* (2008). Basing her work on Baron-Cohen's cognitive empathy deficits view of autism, Barnbaum saw autism as the limit case of full moral agency, where moral judgements are based on either automatically following rules or imitating other people's responses without fully understanding why. Her arguments, if generalised, suggest that Greta Thunberg's environmental activism is either a kind of parroting of genuine moral judgements made by others or that she is not autistic. While it might be unfair to attribute this anachronistic judgement to Barnbaum, Greta Thunberg has recently been accused of both kinds of 'faking' by contemporary critics. Thunberg has replied eloquently to these charges, as I explore below.

While this chapter focuses on autism–empathy–environmental discourses, the purported lack of autistic capacity for moral judgements contributes to the difficulty autistic people have in being believed when they report violence and abuse (see, for example, Dimensions 2019). This urgently needs to be addressed by all autism researchers, both neurodivergent and otherwise. To begin to understand and question the existing discourses on autism, empathy, and environmental experience, I offer a speed-tour of some of the psychological, philosophical, and literary contexts in which they have been addressed, at least in the West. Future work might also consider whether focusing on environmental experience is helpful or if it plays to existing agendas where we are valued only in relation to a neurotypically defined end, such as providing expert knowledge on other species.

As discussed elsewhere in this volume, the philosophical stance of enactivism makes it unlikely that we will find a single neurological basis for autism, even if monotropism offers a helpful guide to a more universal aspect of autistic experience. If the human mind is enactive, it will depend on its social, biological cultural, and material contexts, as well as the life history of the individual. This means that is likely that only part of morality is 'cognitive'; even cognitive psychologists, who arguably would have little to say about the non-cognitive realm of affect and emotions, have asserted that empathy has an affective component. There are intuitively (at least to this author) other ways to experience empathy – corporeal, sensual – which have yet to be investigated (da Silva 2015; Grandin & Richter 2014).

Within the field of autistic life writing, several very high-profile memoirs by autistic authors have engaged with moral issues within environmental and inter-species ethics. The idea that autistics may experience greater environmental empathy may contribute to the 'othering' of neurodivergent people, through the assumption that we are somehow closer to nature than those who consider themselves to be neurotypical. However, this offers fruitful ground for thinking through popular representations of autism, as more people recognise that our times call for new ways of working (that 'business as usual' isn't working). This offers scope for questioning not just what we do, but who does it (even if the eventual gain is for 'normals' rather than all of us). For instance, the young autistic climate activist, Greta Thunberg, states in her memoir *No One is Too Small to Make a Difference* (2019) that her moral clarity is not just possible in spite of, but it's actually due to, her autism:

> I have Asperger's, and to me, almost everything is black or white. I think in many ways that we autistic are the normal ones and the rest of the people are pretty strange. They keep saying that climate change is an existential threat and the most important issue of all. And yet they just carry on as before.
>
> (p. 7)

In line with other discourses that build upon the idea of autistic people having exceptional (if disturbing) skills, Thunberg suggests that autism allows for a kind of moral expertise, and that this is the ability to act upon moral judgements without anticipating recognition and esteem for doing so. As we'll see, this turns the normalising forces of 'recognition' that are so often portrayed as key to *non-autistic* morality, on their head. Thunberg's message works in two ways, according to her audience. If they share with her the assumption that it is possible to be autistic and moral without requiring just one sort of morality (as I believe she suggests) we simply take her claims at face value. If we believe that there has to be only one kind of morality, she may be playfully suggesting to neurotypicals that autistic people have a better claim to being moral, since we are the ones whose behaviour is consistent with our views rather than determined by social norms.

Thunberg's claims to experience moral and epistemic clarity would find very little support from existing medical literature on autism, unless it is accompanied by a kind of rhetorical 'disciplining' that implies there is something socially dangerous about us making moral judgements without external sanction (for more on the way that dominant medical narratives seek to discipline subjects, see Couser, 1997). Like Thunberg, the comedian Hannah Gadsby has described how her autism and reflective 'ability to see patterns' means 'not [having to look] out to the world to see how I should exist' (Valentish & Gadsby, 2018, n.p.).

In line with this, and writing in the *New Yorker* back in 1994, Oliver Sacks affirmed what Uta Frith had said of autistic social 'handicaps', that they have 'a reverse side to this "something," a sort of moral or intellectual intensity or purity, so far removed from the normal as to seem noble, ridiculous, or fearful to the rest of us' (1993/1994, n.p.). The idea of autism as a social handicap perhaps allows us to see some of the ways that the medical model of disability elides its normative model of what it counts to be social.

Yet, rather than appearing ridiculous, in her campaign work in the lead-up to the UN Climate Summit in 2019, Thunberg inspired many autistic and non-autistic activists to join the environmental movement (or to pay heed to her words), and this may even be more likely as a result of her non-normative social identity. Some of this might be down to ableist assumptions regarding the assumption that autistics are 'closer' to nature or moral purity or both, but no doubt it is also due to her intersectional position as a minority youth, neurodivergent, female activist. She exemplifies the possibility of moving from the margins to the centre of global discourse.

As the mock 'Greta Thunberg Helpline for adults angry at a child' shows, she provokes an intense response – hostility, as well as fear and ridicule – especially in 'middle-aged' men (Humphries & Williams, 2019). Yet the many negative responses towards her activism confirm the sinister cultural assumptions about autism, youth, and gender, with autism figuring as the opposite of rhetorically, emotionally functioning humanity, and a subsequent fear that might easily be disguised as righteous anger. For instance, Greta has also been subject to prejudice about autism that is normally saved for autistic adults and other youth who dare to challenge the notion that they might have knowledge that is worth sharing with the world.

While responses to Thunberg's public profile may be compounded by an upsurge in hostility towards minorities in general as a result of right-wing populism, psychologists who noted what they perceive as moral purity in autistic people have failed to explain this perception with any depth. While I do not believe this is a deliberate attempt to dehumanise or scapegoat autistics by psychologists, the suggestion that autistics lack empathy contributes to the othering that amplifies such fears.

Baron-Cohen's writings about autism and absent empathy remain the most influential account, and in its most recent form presents empathy as 'the ability to identify what someone else is thinking or feeling, and to respond to their thoughts and feelings with an appropriate emotion' (Baron-Cohen, 2011, p. 12). While this was originally theorised in connection to a postulated defective ToM in autism, Baron-Cohen now focuses on empathy in relation to purported sex differences: that autistic tendencies towards systematising are a result of our 'extreme' manifestation of the male brain (2003). For Baron-Cohen, systemising and empathising are binary opposites, which are endowed according to gender and neurotype. His recent writing that autism is an 'empathy disorder' implies – as well as other problematic assumptions about gender – both that he believes he is right about what empathy is and that autism is best understood 'from the outside', because self-reports about empathy are misguided.

However, as Sue Fletcher-Watson and Geoff Bird have noted in a recent editorial for *Autism*, 'there is no standard, agreed-upon definition of empathy used in research' (2019, p. 1). Further, 'having the capacity for empathy is often seen as the defining characteristic of being human' (ibid.). The 'use of language that dehumanises [autistic people]' might be connected to 'tragically frequent' 'violations of the human rights of autistic people in residential care services' (p. 5). Fletcher-Watson and Geoff Bird also helpfully summarise the ways in which empathy has been defined in cognitivist debates. While sharing this approach, they are careful to note that autism does not exclude empathy in the ordinary sense. What may alter the emotional response described as affective empathy is a separate condition, called alexithymia but this condition does not preclude Theory of Mind (and, by implication, cognitive empathy) (p. 4).

Fletcher-Watson and Bird suggest there are four main component stages to what is ordinarily considered empathy, rather than the two or three that Baron-Cohen has discussed. These include (A) noticing that someone is feeling something due to their behaviour; (B) correctly interpreting the feeling behind observed behaviour; (C) 'having noticed and correctly interpreted the emotional signals of another person, [the next step] is to feel those feelings – to have an affinity for, resonate with, or mirror – how that person feels' (p. 2). For Fletcher-Watson and Bird, this is what 'we most often refer to when we talk about empathy colloquially' and 'it is also the least easy to measure, potentially the most important, and the only component unique to empathy' (p. 2). Finally, (D) there is the need to decide upon and express a response, and this can lead to misunderstandings since it is possible autistic people are 'not following the same response-script as a neurotypical person' (p. 2).

Autism research can illuminate how a monotropic focus, with a subsequent, although possibly independent, inattention to social cues, and difficulty reading 'across' the autistic/non-autistic divide, may result in neurotypical underestimation of empathy in autistic people along the different stages of this process (pp. 1–2). This is supported by much of the existing autistic life writing. Fletcher-Watson and Bird even suggest that it might be helpful to 'understand the way that empathy might be felt and expressed between two autistic people' (p. 4).[1] I would add that it would be helpful to understand the way that monotropic focus might be felt and expressed between two people, rather than concentrate research exclusively in terms of autistic deviance from a hypothetical norm.

Baron-Cohen has accepted the possibility of intact affective empathy in autistics – defined variously as 'an appropriate emotional response to another person's emotional state' (2003, p. 43) and 'our emotional reactions to people' (2011, p. xi). Yet Baron-Cohen's ability to communicate his theory of autism with a wide audience depends on the elision of these nuances into a single term, without it being explicit that what he most often means by empathy is, in the case of autism, 'cognitive empathy', defined by him as 'the ability to identify what other people are thinking or feeling'. If there is an impairment in autistic people being able to identify non-autistic mental states, this is parallel to the ways in which non-autistics try to understand autistic people, as Damian Milton and others have indicated (Milton, 2012; Chown, 2014). Further, following Fletcher-Watson and Bird, what we ordinarily mean when we talk about empathy is the 'affinity feeling' and this is what people are misled into believing is absent in autism if they are unaware of the wider discussion.

Like Fletcher-Watson and Bird, Baron-Cohen tells a more complex story about how empathy might be diminished in otherwise potentially empathic autistic people when other factors are present (see Baron-Cohen, 2011; Fletcher-Watson and Bird, 2019, p. 4). From this perspective, as well as the enactivist stance mentioned earlier, the idea that empathy defines neurodivergence in general seems particularly questionable.

And yet Baron Cohen and Sally Wheelwright have distinguished a further subtype of affective empathy that should be no more problematic for autistic people than anyone else. They call this sympathy – 'where the observer's emotional response to the distress of another leads the observer to feel a desire to take action to alleviate the other person's suffering' (2004). Empathy in popular discourse also suggests this 'desire to alleviate suffering', rather than the more specific sense of an 'ability to identify what someone else is thinking or feeling' (Baron-Cohen, 2011). While it may be true that autistics and non-autistics struggle to understand other neurotypes, intuitive position-taking is not required in many cases of what Baron-Cohen and Wheelwright call sympathy. It could turn out that sympathy is equally rare in all neurotypes.

The capacity to 'tune in' to other people as required by Baron-Cohen's cognitive empathy, or for steps A and B in Fletcher-Watson and Bird's pathway to empathy, might even hinder other kinds of moral behaviour. When it comes to moral concern for future generations or other species, it becomes clear that, even if we can describe ourselves as *feeling* something like this, we cannot know it. If this feeling depends on a general and non-specific 'desire to alleviate suffering', it might have underpinnings in the 'overwhelming affectual empathy' that some autistic people describe themselves as feeling, alongside a sense of powerlessness about being able to influence the social norms of the present generation.

The denigrated status of autistic people, and our supposed affinity with other species are perhaps factors that initially inspired neurotypical interest in autistic life writing. I believe that there have been, broadly speaking, three 'generations' of autistic life writing in English since 1980, which can loosely be described as:

a   Approximately 1987–1993: those that define or translate what autism is for a non-autistic audience, which were published after the publication of the DSM-III (the first version of the diagnostic manual to include autism in the form of Infantile Autism). These accounts are written chronologically and often build upon, and critique, existing medical representations by describing what it is to live an adult life with autism; and in doing so, lay the foundation for what is to live a good life with autism (even if they may represent the condition as precluding certain aspects of flourishing). These are mainly, if not exclusively, received as narratives of restitution (following Couser, 1997) or 'chaos' in Arthur Frank's sense (1995).

b   1994–2013: those that define a life retrospectively in the context of a later diagnosis of autism for the sake of what earlier experiences contributed to the possibilities of living an ethical life. This generation is influenced both by first-generation works and by the diagnostic criteria for autism in the DSM-IV, which includes Autistic Disorder and Asperger's Disorder which no longer require the onset of 'symptoms' observable by a clinician before 30 months but require the external validation by a caregiver. Coinciding with the autism self-advocacy movement, these works are less inflected by the idea of autism as a pathology or something that precludes selfhood. Writing from the position of their adult life, authors question fundamental assumptions about the nature of autism and need to refer to other autistic people as a source of authority. These are more likely to be read as quest narratives.

c   After 2013: those that seek to intervene in the social world more widely than in cultural understandings of autism. While the DSM-5 continues to define autism in terms of childhood behaviours, these texts name autism as a key aspect of identity (shared by one or more individuals across different age groupings). While they may be received as autoethnography (see Rose, 2008), paratextual discussions of these texts may perpetuate pathological representations of the authors' autism (see McGrath, 2017, pp. 174–176).

Temple Grandin's *Emergence: Labelled Autistic* (with Margaret Scariano, 1986) is an example of the first generation. Dawn Prince-Hughes's (1994) *Songs of the Gorilla Nation: My Journey Through Autism* is a helpful example of the second and Gunilla Gerland's (1997) *A Real Person: Life on the Outside* is ambiguously located between the first and second generations, as both an intervention in broader understandings of autism and as an attempt at 'talking back' from the position of the author's own lived experience/the emerging autism community. It is within this third generation that I locate Greta Thunberg's manifesto/memoir. One of the unique aspects of her work is Thunberg's insight into how the social context of common assumptions about autistics and adolescents will inform her reader's responses to her work.

The general movement in these texts away from medical models of autism, and towards a more socially situated understanding of autism, has happened since autistic life writing was able to reach a wide audience in the 1990s. And yet, each text exceeds this simple classification as it works to construct the narrator who is both recognisably 'a person' and an expert on autism in their own right, in one way or another. While the contradictions and issues involved in this are beyond the scope of this chapter, the emerging autistic discourses about person-hood involve discussions of moral agency that are relevant. It is worth a brief digression into the context in which the texts were received to enable a broader discussion about some of the themes raised.

A brief history of responses to autistic life writing

Early autistic memoirists were criticised in terms of the authenticity of their representations on the basis of their supposed inability to introspect or communicate with an imagined audience (see Sacks, 1993/1994), or if they were granted the ability to introspect and describe authentic experiences, they lacked sufficient ToM to select the sorts of incidents their audience would want to hear about (see Happé, 1991). One prominent idea within literary criticism was that these memoirs could tell us about limits of narrativity and subjectivity, based on assumed medical deficits' in meta-representation and ToM (Jurecic, 2006; Smith, 1996; Zunshine, 2003). Others saw cases like Grandin's as evidence of triumph over a condition that made such writing impossible, or as an exceptional rarity.

Oliver Sacks – a neurologist and writer of a memoir about his own recovery from a mysterious illness – subsequently raised the profile of several autistic life writers, including Grandin, in his essay 'An anthropologist on Mars' (1993/1994). He challenged both humanists and psychologists to reconsider the social and communicative potential development of autistics. Bearing in mind that autism was, at this time, only diagnosed according to supposed developmental differences observed during the first 30 months of a child's life, he lent his professional credibility to the idea that nonverbal infant autism might become highly articulate adult autism. What lay in between remained *terra incognita*.

Second-generation memoirs by autistic writers, which were published after the advent of the DSM-IV in 1994, endorsed a much broader characterisation of autism and Asperger's. Because they no longer required such an early onset of symptoms, these works unsurprisingly present much broader representations of lives and experiences under the label 'autism'. At the same time, the autism self-advocacy and neurodiversity movements were gaining momentum as a result of the work of autistic individuals who understood, and powerfully articulated, how autistic differences in communication and sensory profiles did not preclude *relating* to others as a human being (Sinclair, 1993). The Autism Self-Advocacy Network lent support to first-person accounts of autism through its mantra 'Nothing about us without us'. Self-advocacy and the idea of neurodiversity as a naturally occurring difference supported recognition of autistic moral agency.

Autistic life writing in both print and online forms has experienced huge growth in the past two decades. Many recent works, such as Chris Packham's *Fingers in the Sparkle Jar* (2017) and Greta Thunberg's *No One Is Too Small to Make a Difference* (2019), include descriptions of their authors' ethical beliefs in the widest sense. And yet within literary and rhetorical studies the idea that autistics lack a narrative capacity persists (as Yergeau, 2013, explains). Packham and Thunberg demonstrate that whatever autism is, it is not *defined* by an absence of moral sentiment or narrative and rhetorical skills.

While the neurodiversity movement continues to challenge stigma about autism and other neurodevelopmental conditions, Baron-Cohen continues to describe autism as an 'empathy disorder'. Originally basing his claims about autism and empathy on supposed ToM deficits, from 2003, Baron-Cohen persuaded readers of *The Essential Difference* that autism was an extreme manifestation of a binary 'cognitive' opposition between men and women, with the male and autistic brain capable of systematising only at the expense of the ability to empathise (2003). Inspired by this and debates about the ethics of finding a cure or diagnostic test for autism, Deborah Barnbaum subsequently published *The Ethics of Autism* in 2008. Barnbaum extended Baron-Cohen's argument about empathy deficits to conclude that autistic people are only able to count as moral agents based on rule-following rather than as a result of acting from a (more important) moral feeling or perception. She implied that this afforded some value to autistic lives, but placed fewer obligations on conventional moral agents than the harm that would arise from disregarding autistic subjects from the moral realm. This is because people 'compromise their own moral standing, their own claim to membership in the moral community, when they disqualify others' (p. 102). Once again, autistic morality is represented as 'other' and less important than neurotypical ethical behaviour, and the subjectivity that informs this isn't called into question. Barnbaum's methodological preference for a single moral theory, and unverified supposition of an undeniable non-autistic moral capacity, are called into question below.

While the first generation of autistic life writing written before 1993 broke new ground by positioning autistic writers as authorities on autistic experience once they had 'overcome' the condition through the efforts of others and become 'a person', these texts did not directly address empathy. However, Dawn Prince-Hughes's 2004 memoir *Songs of the Gorilla Nation* described the author's affective empathy and compassion for other species (which built, in some ways, on Grandin's interest in farm animals). This provided the authority that allowed her own claims to be both a moral agent with full personhood, and therefore able to make assertions about her autism. Due to prevailing stereotypes about autism, Prince-Hughes's narrative could still be read as one of 'overcoming' autism.

In her 1996 memoir *A Real Person*, Gunilla Gerland described her desire to lead an ethical life despite being (in her view) both disabled by her autism and by her family circumstances. While Gerland did not seem to consider that autism is compatible with moral behaviour – in fact, her memoir represents a quest to overcome autism for the sake of having the sort of human relationships that are conventionally seen as normal and therefore moral – she also demonstrated how non-autistics fail to achieve meaningful inter-personal relationships.

While first-generation critical autism studies have focused on these 'from the inside' accounts and tried to translate them into recognisable experiences for neurotypical audiences (Davidson & Smith, 2009; Solomon, 2010 and 2015), there has been limited attention paid to descriptions of ethical sentiments in autistic life writing, let alone willingness to assume that they might tell us anything worth knowing about individual lives. Those who are exploring 'autistic' forms of rhetoric and language helpfully identified how first-generation writers such as Grandin and Prince-Hughes are subject to the pressure to translate their writing into work that meets the expectations of non-autistic readers, for instance in the use of language and in the requirement for disclosure (see Rodas, 2018, pp. 21–23; Murray, 2008, p. 33). These texts employed, to various extents, recognised 'discourse conventions' (Yergeau, 2018, p. 21), and succumbed to 'market demands' since they were '[g]rounded in the heroic tradition of the *Bildungsroman*, or the traditional overcoming narrative, confession or apologia' (Rodas, 2018, p. 21). Yet to focus on this exclusively fails to do justice to the ways in which *any writer* is confined by their knowledge of existing literary conventions. The life narratives of Prince-Hughes, Gerland, and Thunberg may indeed be read as *autistic testimonio*, since, as Irene Rose has observed, they offer a 'recounting of group oppression' and demand 'an active reader response' (Rose, 2008, p. 48). As a manifesto for an audience that is assumed to share the same response, Thunberg's work may be read as both autistic and youth-environmentalist *testimonio*, and as an attempt to name autistic moral agency outside of the *Bildungsroman* tradition.

Empathy across neurotype and species

As noted above, *Songs of the Gorilla Nation* (2004) is Dawn Prince-Hughes's memoir of her early life and her adults diagnosis of Asperger's at a time when she also discovers her vocation (and as such is a *Bildungsroman*); but it is also a work that situates her autism as both 'like and unlike' other people's autism, and she refers readers to works by Grandin and David Miedzianik (Rose, 2008, p. 48). Gunilla Gerland, in *A Real Person: Life on the Outside* (1997), is similarly concerned with her own spiritual growth, but she also challenges conventional ideas about autism in Sweden at the time of writing. Greta Thunberg's memoir/manifesto *No One Is Too Small to Make a Difference*, a generation later, witnesses the author's struggles to gain recognition as a moral agent in the context of both her autism and the climate crisis, and her work arguably speaks to anyone who is struggling to influence anthropocentric behaviour, regardless of neurotype.

Prince-Hughes's narrative encompasses her turbulent childhood and adolescence. She described her own social struggles, her affinities with the natural environment and early experiences of her sexuality. The 'coming of age' aspect of her account did not involve normalisation or overcoming any of her 'queer' tendencies, but she linked her own traumatic experiences to oppression faced by others. However, her authority as a witness to events may be constructed either on the basis of later expertise as a scientist or in relation to her role as an autoethnographer creating a 'collective record' of the ways autistic voices have been oppressed (Rose, 2008, p. 47).

Like all the authors considered here, Prince-Hughes described her lifelong desire for moral purpose, for meaning defined as connection with human others, and for 'companionship that validates one's experiences from afar' (Prince-Hughes, 1994, p. 33). While understanding that both these latter were at odds with popular understandings of autism at the time (which were based on ToM deficiencies), she urged an understanding of 'direct sources of experience' of autism, since this helps to overcome over-generalisations based on 'known patterns of autism' (p. 7) and a limited number of examples.

Prince-Hughes's narrative climax centres on her reconfigured understanding of the social world. After a period in which she began to observe a family of gorillas at a Seattle zoo, she started to see her own life differently. As a result of her supposed social difference, she began to compare herself to both the captive gorillas and humans 'who are not bright on the stage of common action' (p. 4). She found in the literal glass that separated the observers from the gorillas a symbol for the boundary between the neurotypical gaze on the human or animal other. While earlier authors had described themselves as other, Prince-Hughes posited her own, and the gorillas', difference as produced by the mechanisms that were designed to facilitate their interaction – the zoo. And like the glass barrier that separated the gorillas from their human observers, the gaze can be both metaphorically and literally interrupted or broken.

Prince-Hughes described her interactions with a male gorilla called Congo. She retrospectively narrates the experience of feeding him strawberries as the first time 'she connected to a living person' as she 'never had before'. Laying fruit at the edge of the enclosure, between the bars and the glass, Prince-Hughes is 'compelled to put the berries in the same repeating order', which results in Congo and Prince-Hughes putting their 'fingers down at the same time'. Congo's

> gigantic finger, black and leathery, soft and warm, rested on my own digit. We stared at our fingers, neither of us moved. Finally, I looked up into his soft brown eyes. They were dancing with surprise.
>
> (p. 6)

The significance of this encounter, for Prince-Hughes, is that she finds a reflection of her own urges for repetition and ritual and a sense of 'what it is to not be alone' (p. 6). She imaginatively placed herself in the position of Congo and attempted to reverse the direction of the gaze. Reflecting on the ritualistic aspects of such play in the gorillas, Prince-Hughes noted that it may have another function in both humans and gorillas:

I began to understand ritual and its power a bit more. I had the advantage of watching my gorilla family in ritual activity, sometimes as a reaction to their confinement but often born of a spiritual, an aesthetic, even an educational need. At this time, I learned the value and beauty of ritual.

(p. 19)

While admitting that the gorillas' repetitive behaviours may be a response to the restrictive conditions of the zoo, Prince-Hughes suggested that rituals, repetitions, or 'perseverances' for autistic people, may provide sources of pleasure. Through this and other examples of her own sense of affinity with – and other people's dis-affinity with – the gorillas, Prince-Hughes is motivated to pursue a career in gorilla conservation. While Temple Grandin described herself as an anthropologist on Mars, Prince-Hughes presented herself as a xenobiologist presenting the 'normal human' as other.

Morally ambivalent empathy: the pain caused by assumed cognitive empathy

Gunilla Gerland described her early life, prior to diagnosis, in her 1997 memoir, *A Real Person*. While unhappy with the 'high functioning' classification of her autism diagnosis, since it 'sounded like something you might say about an object that was slightly defective' (1997, p. 239), it allowed her to think of her difference having a biological basis rather than a moral origin, and it allowed her to understand herself as a 'real person' rather than one who was deliberately difficult, defective, or lazy (p. 238). While she generally reiterated a pathological view of autism as a handicap, she did not seem to think that this prevented her from being morally concerned for other people (particularly her sister, Kerstin).

Gerland's spiritual journey was, like Prince-Hughes's and Grandin's, one that depended on 'overcoming' of social limitations. However, like Prince-Hughes and Hannah Gadsby, she did not consider herself to need social recognition to authorise her own version of events or to form judgements about others, even as a child. Although Gerland's childhood and adolescence were marked by both emotional and physical abuse at the hands of her father, and later by her mother's alcohol and drug use, her memoir was chosen by both Barnbaum and Baron-Cohen as an example of autistic empathic failings. Discussing the possibility of autistic ethics based on rule-following in 2009, Baron-Cohen, who presumably had not read Gerland's memoir, repeated Deborah Barnbaum's comments about Gerland in *The Ethics of Autism: Among Us, But Not of Us*.

Gunilla Gerland, who has autism and describes how she was unperturbed by the death of her father, comparing his loss to a bowl of fruit that was on the table one day and gone the next.

(Baron-Cohen, 2008)

While he concludes that removing autistic people from the 'moral community' would be immoral, Baron-Cohen considers Gerland's writing an example of the solipsism that precludes the 'visceral' response that ordinarily produces moral action. However, the suggestion that Gerland's father had died is a misunderstanding: he had simply moved out. Gerland's responses throughout *A Real Person* are extremely visceral, and this is why it stands out as an exceptional piece of writing. Like Barnbaum, he omitted from mentioning that Gerland was only an infant at the time and that her father had also been abusive (Gerland,1997, pp. 42–43).

While Barnbaum refers to the fact that Gerland's father had only moved out and had not died, she does not connect this to Gerland's early difficulties making predictions about the future, nor does she mention Gerland's early concern for her sister's wellbeing. This suggests that Barnbaum had read *A Real Person* with the intention of finding instances of empathic deficiencies in accordance with her earlier reading of Baron-Cohen. Barnbaum's account exemplifies the self-fulfilling prophecy of the neurotypical gaze on an autistic subject.

Gerland became an autism advocate after publishing this memoir, working to educate professionals on how to engage more compassionately with verbal and non-verbal autistic people. She also became one of the pioneers of autistic participation in research on autism (see, e.g. Gerland, 1997). As her work, like Grandin's, came with recommendations from the prominent clinical psychologist, Christopher Gillberg, we may assume that the neurotypical gaze may have shaped the kinds of stories Gerland told about both her own and collective autism. And yet, neurotypical intervention may have provided an opportunity for Gerland's individual self-, and self–other-, reflection. Gerland contrasted her own biological understanding of her autism with what was then the conventional psychoanalytical view that autism resulted from deficient parenting – in fact, she turned this view on its head. She stated that her autism helped her avoid becoming too 'neurotic' as a result of that same bad parenting (1997, p. 250). The resulting story is indeed one of triumph over the adverse conditions of a 'biological handicap' and a 'dysfunctional family' (p. 250). Yet in Gerland's description, neither handicap nor dysfunctional family preclude her from having experiences which, according to Fletcher-Watson and Bird, are what is ordinarily meant when we talk about empathy.

Talking back: autism as moral motivation

While Gerland and Prince-Hughes describe their moral feelings, Greta Thunberg's manifesto *No One Is Too Small to Make a Difference* requires us to take the possibility of her ability to make moral judgements as a given, so we are then able to critique the mere suggestion that autism can be defined as lack or deficiency. While Gerland and Prince-Hughes talk back to standard depictions of autistic empathy deficits and cast cognitive empathy as either problematic or unnecessary, Greta Thunberg (playfully) suggests that to lack cognitive empathy may actually, in some cases be a moral virtue. Gerland's (presumably neurotypical) mother is represented as lacking enough empathy to know that her daughter hates birthday parties (1997, p. 41). Prince-Hughes emphasises her own affective empathy for those whose mental states she cannot fully access, including those belonging to other species. Each author tells us something different about the impossibility of identifying any individual 'faculty' that will produce morally optimal outcomes in all cases. The possibility of autistic concern for other species offers a chance to 'reverse' the assumption that cognitive empathy is essential to moral behaviour, and to turn the gaze towards what might be missing in 'neurotypical' morality. Thunberg confronts us with the possibility that an unnamed group of cognitive others – future humans – depend on those who are motivated to act without typical social recognition, because they have had to find other ways to exist in a world that sees them as having less value.

Yet her demand for radical changes to society to prevent climate change has been met with criticism that echoes the denial of autistic empathy on the basis that it does not conform to neurotypical empathy. Andrew Bolt, who is a broadcaster on Australia's Sky News, linked her claims to an alleged underlying pathology:

> She suffered years of depression and anxiety attacks and was finally diagnosed with Asperger's syndrome, high-functioning autism, and Obsessive-Compulsive Disorder. Her intense fear of the climate is not surprising from someone with disorders which intensify fears.
>
> (Bolt, 2019)

If Bolt had taken the trouble to register what Thunberg had actually said about her autism, he'd need to respond differently. She states that it was not anxiety or compulsions that drove her actions to raise awareness about the threat of climate change, but her autism itself. In *No One Is Too Small to Make A Difference*, she says of her autism:

> Some people mock me for my diagnosis. But Asperger is not a disease, it is a gift. People also say that since I have Asperger I couldn't possibly put myself in this position. But that's exactly why I did this. Because if I would have been 'normal' and social I would have organized myself in an organization or started an organization by myself [...]. But since I am not that good at socializing I did this instead. I was so   frustrated that nothing was being done about the climate crisis, and I felt like I had to do something, anything.
>
> (2019. p. 30)

Thunberg, here, casts standard rhetoric about autism defined by social deficits on its head. She implies that if she had placed a greater value on conformity with her peers – if she had a tendency to pick up on social cues or found herself 'naturally and spontaneously tuning into someone else's thoughts and feelings, whatever these might be' (p. 21) – she would have found another (and possibly less effective) way to campaign to reduce global carbon emissions. Perhaps being relatively *more* tuned in to those who share similar assumptions, or relatively more tuned in to the environment, through a monotropic focus, might have helped her get the message across.

However, I also believe Thunberg is also here knowingly performing, and thus parodying, the idea of autistic moral 'purity' that Frith (2014) describes: the work, as a whole, resists the idea that the narrator is superior to her assumed audience. Rather than saying that her autism limits her or gives her superpowers, she suggests that it has simply become a condition that has produced this particular outcome. And in this way Thunberg gestures towards a new understanding of autism – as simply difficulties that occur in some situations, rather than as a condition defined by moral limitations. Unlike Grandin or Prince-Hughes, her authority does not depend on any assumed ability to 'speak for nature', but it perhaps depends on the emergence of the voice of autistic adolescence – one that had been assumed not to exist.

The idea that autistic people are unable to make moral judgements, or are only able to blindly follow rules, speaks mainly to a normative urge to find a single story about what makes a good life. Meanwhile, the same story deprives us of the essential agency that is necessary for us – and possibly all of humanity – to flourish. The story of autism as defined by empathy deficits also plays to totalitarian conceptions of the good, since the world in which we live is dependent on multiple visions of what is right. Since even when recognised, autistic moral agency risks being co-opted into utilitarian enframings, it needs to be rearticulated.

In fact, to cast any neurotype as inherently pathological or valuable creates a situation in which groups who are perceived to share that trait are at risk of being sacrificed for the greater good. When we seek to locate a single feature such as empathy as a unique sign of our supposed individual worth, we are also at risk, not of debasing ourselves, but of not recognising our ongoing need to refine our own judgements according to the new circumstances in which we find ourselves.

This work was supported by the Wellcome Trust [218124/Z/19/Z]

Acknowledgements: with thanks to Nick Chown, Hanna Bertilsdotter Rosqvist, Robert Chapman, and David Perkins for their invaluable suggestions on this chapter.

Note

1 However, it seems to me that there may be a contradiction between the idea that empathy is misunderstood and the quest to address the 'paucity of cognitive models of empathy' (Fletcher-Watson & Bird, p. 4) given that empathy is in their own definition *affectual* as well as cognitive.

References

American Psychiatric Association (2013). *Diagnostic and statistical manual of mental disorders* (5th ed.). Washington, DC. American Psychiatric Association.

Barnbaum, Deborah (2008). *'Among us, but not of us': The ethics of autism.* Bloomington, IN: Indiana University Press.

Baron-Cohen, Simon (2003). *The essential difference: Men, women and the extreme male brain.* London: Allen Lane.

Baron-Cohen, S. (2009). Does autism need a cure? *The Lancet, 373*(9675), 1595–1596.

Baron-Cohen, Simon (2011). *Zero Degrees of Empathy.* London: Penguin.

Baron-Cohen, Simon, & Sally Wheelwright (2004). The empathy quotient: An investigation of adults with Asperger syndrome or high functioning autism, and normal sex differences. *Journal of Autism and Developmental Disorders, 34*:163–175.

Bolt, A. (2019). The disturbing secret to the cult of Greta Thunberg. *Herald Sun,* 1 August.

Chown, N. (2014). More on the ontological status of autism and double empathy. *Disability and Society, 29*(10): 1672–1676.

Couser, G. Thomas. (1997). *Recovering bodies: Illness, disability, and life writing.* Madison, WI: University of Wisconsin Press.

da Silva, A. A. (2015). Researching empathy through staged performance. (Unpublished dissertation). Design School Kolding.

Davidson, J., & Smith, M. (2009). Autistic autobiographies and more-than-human emotional geographies. *Environment and Planning D: Society and Space, 27*(5), 898–916.

Dimensions (2019). Why don't people understand? Online at www.dimensions-uk.org/get-involved/campaigns/say-no-autism-learning-disability-hate-crime-imwithsam/know-more/dont-people-understand/ (accessed 24 January 2020).

Figueroa, Robert Melchior Figueroa (2017). Autism and environmental identity: Environmental justice and the chains of empathy. In Sarah Jaquette Ray & Ray Sibara (Eds.), *Disability studies and the environmental humanities.* Lincoln, NE: University of Nebraska Press.

Fletcher-Watson, S., & Bird, G. (2019). Autism and empathy: What are the real links? *Autism, 24*(1).

Frank, Arthur (1995). *The wounded storyteller.* Chicago, IL: The University of Chicago Press.

Frith, Uta (2014). Autism: Are we any closer to explaining the enigma? *The Psychologist, 27*(10):744–745.

Gerland, Gunilla (1997). *A real person: Life on the outside* (translated by Joan Tate). London: Souvenir Press.

Grandin, Temple & Margaret Scariano (1986). *Emergence: Labelled autistic.* Ann Arbor, MI: Ann Arbor Publishers.

Grandin, Temple & Richter, Ruthann (2014). 5 Questions: Temple Grandin discusses autism, animal communication. Online at https://med.stanford.edu/news/all-news/2014/11/5-questions-temple-grandin-discusses-autism-animal-communicati.html (accessed 24 January 2020).

Happé, Francesca (1991). Autism and Asperger syndrome: The autobiographical writings of three Asperger syndrome adults: problems of interpretation and implications for theory. In Uta Frith (ed.), *Autism and Asperger Syndrome* (pp. 207–242). Cambridge: Cambridge University Press.

Humphries, Mark & Evan Williams (2019). Mark Humphries presents the Greta Thunberg helpline. www.abc.net.au/7.30/mark-humphries-presents-the-greta-thunberg-helpline/11561522 (accessed 14 November 2019).

Jurecic, Ann (2006). Mindblindness: Autism, writing, and the problem of empathy. *Literature and Medicine, 25*(1), 1–23.

McDonagh, Patrick (2013). Autism in an age of empathy: a cautionary critique. In Joyce Davidson and Michael Orsini (Eds.), *Worlds of autism: Across the spectrum of neurological difference*. Minneapolis, MI: University of Minnesota Press.

McGrath, James (2017). *Naming adult autism*. London: Rowman & Littlefield.

Milton, D. E (2012). On the ontological status of autism: The 'double empathy problem'. *Disability & Society, 27*(6):883–887.

Murray, Simon (2008). *Representing autism: Culture, nature, fascination*. Liverpool: Liverpool University Press.

Packham, Chris (2017). *Fingers in the sparkle jar*, London: Ebury Press.

Prince-Hughes, Dawn (1994). *Songs of the gorilla nation: My journey through autism*. New York: Harmony.

Rodas, Julia Miele (2018). *Autistic disturbances: Theorizing autism poetics from the DSM to Robinson Crusoe*. Ann Arbor, MI: University of Michigan Press.

Rose, Irene (2008). Autistic autobiography or autistic life narrative? *Journal of Literary Disability, 2*(1), 44–54.

Sacks, Oliver (1993–1994). An anthropologist on Mars. *The New Yorker*, 27 December 1993 and 3 January 1994. www.newyorker.com/magazine/1993/12/27/anthropologist-mars (accessed 13 September 2019).

Sinclair, J. (1993). Don't mourn for us. *Our Voice*, 1(3). URL: (Retrieved December 2019) www.autreat.com/dont_mourn.html

Smith, Sidonie (1996). Taking it to the limit one more time. In Sidonie Smith & Julia Watson (Eds.), *Getting a life: Everyday uses of autobiography*. Minneapolis, MI: University of Minnesota Press.

Solomon, O. (2010). What a dog can do: Children with autism and therapy dogs in social interaction. *Ethos, 38*(1), 143–166.

Solomon, O. (2015). 'But-he'll fall!': Children with autism, interspecies intersubjectivity, and the problem of 'being social'. Culture, Medicine and Psychiatry, 39(2), 323–344. doi: 10.1007/s11013-015-9446-7.

Thunberg, Greta (2019). *No One is Too Small to Make a Difference*. London: Penguin.

Valentish, Jenny & Hannah Gadsby (2018). 'I broke the contract': how Hannah Gadsby's trauma transformed comedy. *Guardian*, 16 July 2018.

Yergeau, Melanie (2013). 'Clinically significant disturbance: On theorists who theorize theory of mind. *Disability Studies Quarterly, 33*(4). http://dsq-sds.org/article/view/3876/3405 (accessed 26 September 2019).

Yergeau, Melanie (2018). *Authoring autism: On rhetoric and neurological queerness*. Durham, NC: Duke University Press.

Zunshine, Lisa (2003). Theory of mind and experimental representations of fictional consciousness. *Narrative, 11*(3): 270–91.

Chapter 8

Sensory strangers

Travels in normate sensory worlds

*David Jackson-Perry, Hanna Bertilsdotter Rosqvist,
Jenn Layton Annable, and Marianthi Kourti*

Introduction

In this chapter we interrogate the normative pressures that are brought to bear
upon sensory experiences that are considered to deviate from the ordinary, to
seek out the political importance of the senses. Leaning on queer theorist Alison
Kafer's definition of 'political' (2003, p. 9) we propose that sensory experiences –
and our possibilities to discuss them – are 'implicated in relations of power and
that those relations, their assumptions, and their effects are contested and contest-
able, open to dissent and debate'. The senses have been sporadically theorised and
studied within sociology; however, while Georg Simmel coined the phrase 'sociology of the senses' in 1908, over a century later a formal sociology of the senses
is 'a field so new that it barely even exists' (Vannini, Waskul, & Gottschalk, 2014,
p. 11). Furthermore, the field has generally been more concerned with studying
the senses themselves and relationships between senses and culture than with considering the senses as a possible site of political pressure (cf. Rhys-Taylor, 2013,
p. 364). Here, we lean on Georg Simmel's theorising of The Stranger to propose
a reading within which the 'sensory stranger' provides a valuable epistemic asset.
This approach echoes critical autism studies, whereby autism may be conceptualised not just as a (largely pathologised) 'object' to be studied, but 'as a vantage
point from which the range of humanity can be viewed' (Murray, 2012, p. 36).

Autistic people's sensory experiences are often pathologised, considered by
diagnostic definition to be abnormal, and by anecdote to be extraordinary or other-
worldly (Bogdashina, 2016). Considering them otherwise, as we do here, is valuable on several counts. An exploration of autistic bodyminds which does *not* take
as its point of departure an assumption of pathology, reflects potentially creative 'discrepant sensorialities' (Serlin, 2017) that trouble sensory hierarchies and
acknowledge multiple co-existing forms of sensory and embodied perception, in
which the impact of language is ever-present. Focusing on the first-hand sensory
experiences of autistic people can also recontextualise autism, shifting emphasis
from a model of deficit to a model of neurodiversity. This shift creates an alternative, empowering framework in relation to which autistic people can position
themselves. 'Autism', as Murray (2012, p. xiii) notes, 'is frequently talked about,

but it is rarely listened to'. This is to say that 'the vast majority of research in autism is still undertaken on autistic people, rather than *with* them' (Chown et al., 2017, p. 1). Here, and within neurodiversity studies more broadly, we seek to add to the epistemological and ethical case for 'listening to' autistic experience but also the imperative of neurodivergent 'talking back' to a predominant deficit discourse, and through this 'talking back', to challenge the silencing of autistic voices in the fields in which autism is studied (Milton, 2014).

The chapter is based on a cross-neuro-status collaboration, which we see as an important part of a process of co-creating knowledge, meaning, and methods to contribute to a deeper understanding of both autistic experience and the world in which the lives of people of all neuro-types unfold. While we refer to categories such as 'neurotypical', 'autistic', 'neurodivergent' and so on, we consider these positions to some extent as 'performative acts' (Butler, 1990). Put differently, the sensory experiences of a neurotypical individual may be perceived as having distinct features from that of a neurodivergent one, but it is also a position that is discursively and thus politically produced (cf. Nadesan, 2005). Thus '(w)hat we (think we) know about autism – the so-called facts – are difficult to dislodge or disentangle from the social and discursive worlds in which they are embedded' (Davidson & Orsini, 2013, p. 14). New ways of describing (and sharing) sensory experience challenge societal attitudes and barriers that cause discomfort for individual neurodivergent people.

Theoretical positioning

We are involved here in a 'cripistemological' undertaking. A cripistemological approach involves producing 'first-hand, and in some cases, first-person knowledge about topics that concern disabled people and communities, broadly conceived' (McRuer & Johnson, 2014, p. 158). 'Broadly conceived' in this quote is important for our purposes. Not all autistic people identify as being disabled or are comfortable with an approach that positions them as disabled people; others do and are. References here and later to disability theorists are therefore *not* a statement of ontological positioning (that autism 'is' a disability, and that autistic people 'are' disabled), but an acknowledgement of two points. First, of the *usefulness* of some strands of disability theory in approaching autism, and second, of our opinion that autistic people and communities are subject to many tensions in common with disabled communities. We consider this notably, here, in relation to questions of (lack of) epistemic authority and discursive power relations (see Bertilsdotter Rosqvist & Jackson-Perry, 2020).

A cripistemological approach situates knowledge production within discursive systems of power, and privilege (Patsavas, 2014). At a time when 'we know more about autism than we've ever known, what we know is very little, and what we know is decidedly non-autistic' (Yergeau 2018, p. 11). From the perspective of the neurodiversity paradigm, situating knowledge production within discursive systems of power and privilege (cf. Patsavas, 2014) means to *centre* neurodivergent sensory experience. This necessarily requires communication on our own terms in spaces of our own, which in our group could be referred to as a neurodiverse 'epistemological community' (Nelson, 1993).

To illuminate ways in which discursive systems operate, it is necessary 'to open up spaces of translation between researchers and others whose experiences might otherwise be excluded from the field of sensory studies' (Rhys-Taylor, 2013, p. 364). This 'translation' is undertaken here by a group of researchers of whom the majority are autistic. We use the analogy of The Stranger, theorised by Simmel as a *spatial* outsider, as someone arriving in a new group from a distinct culture, a person 'determined, essentially, by the fact that he [*sic*] has not belonged to it from the beginning, that he imports qualities into it which do not and cannot stem from the group itself'[1] (Simmel & Wolff, 1950, p. 402).

It is through this distinctive subjectivity that '(the stranger) becomes essentially the man [*sic*] who has to place in question nearly everything that seems to be unquestionable to the members of the approached group' (Schütz, 1944, p. 502). We do not take this questioning of 'nearly everything' to mean that nearly everything *separates* people of different neurotypes or that the stranger is 'far' from other members of the group: on the contrary, the stranger 'is an element of the group itself', and therefore strangerhood is 'a specific form of interaction' (Simmel & Wolff, 1950, p. 402). Thus, the stranger and the non-strangers are not necessarily fundamentally different, but rather the qualities inherent in the stranger's position, here the neurodivergent, make possible a questioning of the norms of the culture that surrounds them. Not only will the stranger be able to question the generally unquestionable, but through the stranger's status 'through his [sic] distance from the common experience of the group in which he finds himself' the stranger 'is freer practically and theoretically; he surveys conditions with less prejudice; his criteria for them are more general and more objective ideals; he is not tied down in his action by habit, piety, and precedent' (Simmel & Wolff, 1950, p. 405).

However, strangers *may* be tied down by the ways in which the broader group *describes* them (cf. Nadesan, 2005), and indeed the discursive possibilities to describe *themselves* that may be foreclosed by the description of the majority group. The dominant reading of autistic people in the scientific and professional literature, despite increasing critique, is still overwhelmingly that of deficit (Dinishak, 2016). This may limit the possibilities that autistic people may have of considering themselves *other* than through this lens.

In the case of neurodivergent sensory experience, the very element that potentially renders the stranger's experience invalid (within the paradigm of pathology), the representation of their experience as being outside the norm, as being deviant, also uniquely positions that same person to question and jostle that very norm: to travel otherwise through normate sensory worlds. We explore this 'travelling otherwise', and in doing so question the 'habit, piety, and precedent' that may exert discursive and political pressure on sensory subjectivity, thereby calling 'attention to the invisible privileges of normative practice' (Serlin, 2017).

Methods

We are variously situated geographically, neurologically, and in our academic disciplines and levels of experience. Here, we have used our own collective autoethnographic data, our sensory stories, as the focal point from which to examine and interrogate sensory norms. Following an initial day-long meeting of all the authors, we used what we term 'sensory story-telling' as a form of data collection and analysis. Broadly, this involved an iterative process whereby one of the authors started writing a free text, with no 'guiding' beyond 'sensory experience'. In this way we sought to enable the person to 'travel' where they wanted around the theme rather than walk a path determined by pre-existing ideas of the group, although during the initial day-long meeting the theme of sexuality had been identified and discussed in some detail. This form of data production was then carried out by the other contributing authors, initially producing three texts. These texts were circulated, added to, and commented upon by other authors during 'virtual' group writing rounds, two residential in-person writing retreats attended by two contributors, and online co-writing sessions. The creation and analysis of data thus became a form of intertextual intimacy. In this an ongoing dialogue between author-analysers led both to an iterative development of 'writing up results', and a development and refining of sensory stories – and their analysis – throughout the process.

Language: limits and possibilities

This process produced a considerable quantity of data. Much of our data reflected an awareness of the insufficiency of language, and the limits and possibilities of re-imagining neurodiverse sensory experiences within sensory normate language, that is to say, generally medical or clinical narratives of sensory experiences, which either ignore the existence of sensory divergent differences or else pathologise them. We focus here on parts of our data that illustrate how authors move away from normative or pathologising discourses to alternative understandings of their sensory experiences.

We can hear echoes of Simmel's stranger, who lives *within*, but does not necessarily identify as being *of* a 'host' culture, in the following:

> I often feel that everyone else has the same cultural background [to start from] and I don't. And that makes me feel broken, out of place, less, someone who cannot be accounted for, included or accommodated for.

Further, we are here discussing an area that contains assumptions of normate sensory processing and expression. This is clearly not simple for the stranger interfacing with their 'host' group, and may also be also a difficulty *between* members of such a heterogeneous group as 'autistic people'. In our research collective we found numerous examples of differences that could impact various stages of this project, that potentially overrode neurostatus. This included factors such as differing first languages, native cultures, and sexual/sensory experiences. Teasing out which differences were due to neurostatus and which to other factors largely depended, here, on the ways in which the affected individual identified these differences. We embarked on our process through considering the limits and possibilities of language, redefining and formulating new concepts through our conversations within our group. In the following quote we see an expression of the limits of language:

> … when I am stroking my cat, I cannot tell you how that feels or why; I can just tell you that I do it and you are left to put the rest of the pieces of what that means yourself.

The singularity of experience illustrates a difficulty of expressing 'senso-reality experiences' (cf. Serlin, 2017) which are not easily approached through dominant language:

> Autistic people with learning difficulties … may have a wealth of senso-reality experiences that just cannot be approached through language, […], because the senses are something we don't usually talk about to a great extent and so we are very limited discursively.

This complexity was present implicitly throughout our own writing process, and touched on during one of our exchanges:

> … [this experience] is really well described and I totally get it but it is something totally alien to my own experience. How are we [as a group, and wider culture] supposed to know if we have similar experiences if the only medium to communicate them with is through language which is flawed as we have seen. I could never have described [this] experience with language so I would ask us to reconsider the language we have used here.

The limits of language are further developed in the next conversation. In the following quote we see how, contrary perhaps to the assumption of a common 'us', the stranger carries an expectation of singularity – where an 'invented' language of one's own, a 'neurodivergent mother tongue', is brought to express one's experiences on one's own terms. This brings other difficulties, as one of us says in describing communication with a partner:

> how on earth are you supposed to communicate about sex in multi-lingual-autistic-cat-anime with someone who only speaks English-cat and is now learning autistic.

This invented language of one's own, 'multi-lingual-autistic-cat-anime', may be shared by just a few, or maybe only one other person and, even then, only partially:

> It's frustrating when only one person in the world speaks your language and everybody else can only communicate with you through your second, third, fourth etc. language but not your mother tongue.

Another example of the difficulties of dominant language to describe one's sensory experiences could be appropriating visual language while also finding it too difficult to make it shareable with another. One of us describes how they incorporate nonverbal communication in their language, for example, anime:

> ... the way [the author] portrays sensory experiences has given me almost some form of late-onset synaesthesia ... [they] turn into pictures and sounds ... they created a vocabulary in my head that is in pictures and sounds (not words) that are attached to sensory experiences. If you ask me how my body feels, the only thing that comes to my head is this [see Figure 8.1]:

Figure 8.1

On the other hand, they may find it too difficult to make the translation from this language of one's own into a shareable language, and choose to remain silent:

> Because I have no other way to explain to you how my body ripples touching each other feels for me and that I don't like it. But it takes a long time to go from a picture to words, so I just don't talk about it.

Synaesthesia (whereby information generally perceived and/or expressed via one sensory or cognitive pathway is experienced via another pathway) is more common among autistic than non-autistic people (Ward et al., 2017). Furthermore, some autistic people are multiply neurodivergent, which might make finding a shared language even harder. While this invented language of one's own works as a way of pointing out the limits of sensory normate language, it may result in silence, a sense of resignation, an experience of a non-shareable language about one's sensory experiences in social contexts.

However, in different ways, two of our writers sought out alternatives to the normative language of diagnosis by drawing on their sensory experiences, one through accessing online autistic communities and the other seeking out fulfilling sensory experiences via a BDSM[2] lifestyle and sensory integration therapy. In this way, these writers felt that they emancipated themselves from the stigmatisation that diagnosis may inflict. This was not a simple thing to do, and it took both emotional labour and access to specific communities and languages, both online and 'in real life'. This brought with it a similar tension concerning hierarchies of knowledge. For example, biomedical language of diagnosis has at least initially a high level of epistemic authority (Oikkonen (2013, p. 284). In relation to that more dominating language either the 'language of one's own' or community languages that have been developed in certain alternative epistemological communities, such as the autistic community (Belek, 2019) or affirmative professional communities, may find it hard to gain credibility.

Indeed, diagnosis, and the biomedical narratives on which it leans, can be both 'a tool and a weapon … a source of knowledge, sometimes trustworthy and other times suspect … shaped by particular belief systems, useful and dangerous by turns' (Clare, 2017, p. 41). Here we see echoes of Garland-Thomson's (2005, p. 1567) rewriting of 'oppressive social scripts' in order to reimagine a disabled self. She describes ways in which the biomedical narrative 'casts the variations we think of as impairment as physiological failures or flaws'. A response to this can be to formulate 'a logic that allows people to claim the identity of disabled without having to conceive of it as a diminishment of the self'. This is increasingly the case in emergent research carried out in neurodivergent circles (Chown et al., 2017), which is both a product of, and feeds back into, autistic communities as well as communities of affirmative professionals. The diagnosis of autism is conceptualised as a recognition of a way of being in the world that, while it may well carry greater or lesser degrees of disadvantage, also holds emancipatory and explanatory value for individuals:

> Finding out that out [that I was autistic] gave me the language to describe many of my experiences, which I am very grateful for.

In this way, a sensory self initially arising from a biomedical deficit model as being marked with sensory deficits, is strategically reimagined as a sensorially different self, intelligible in its own right (Belek, 2019), but not without struggles for rights of interpretation. The following expresses this struggle, but also a sense of triumph: community language or affirmative professional language works as an enabler of the expression of alternative sensory experiences, or the possibilities of 'owning the words that describe my own experiences':

> This was an interesting process for me, having originally been exposed to autism theory in academic settings as a part of my Special Educational Needs undergraduate programme for four years, only to discover my own autistic identity after a few hours of watching YouTube videos of autistic people talking about their experiences. This initiated a complex navigation of epistemological locations, between the online autistic groups and the professional books and articles I was familiar with, the negotiation of which was the closest approximation [I had] to my [own] experiences and, indeed, the 'entity' of autism.

This comment reflects the experience of epistemological travelling; from the location of an outsider within an autistic community, rediscovering themselves and their own autism with the support of autistic community language. As Belek (2019, p. 39) puts it,

> (O)btaining an autism diagnosis leads autistic adults to engage with other members of autistic communities and to become exposed to such relevant systems of knowledge as the language they use, the ideologies they subscribe to, and the discourses they produce.

Going from 'being talked about' to being an active participant in the conversation about their own autistic subjectivity, or what we refer to as 'talking back', held powerful possibilities to rethink their life and the agency they felt they could wield:

> Figuring out little by little not only what my relationship with myself was, but also contextualizing my relationships with others, the ways in which others storied me (e.g. as someone who is 'talented' in their work with autistic students). It became a powerful source of agency in my personal life, whether I felt able or empowered to exercise it or not. [It became] an epistemological and discursive location that gave me the freedom to approach my life phenomenologically from the events, rather than theoretically from abstract,

and very flawed, theoretical frameworks I was familiar with. Ultimately [it empowered] me in the process of owning the words that describe my own experiences.

Similarly, through the language of sensory integration assessment within an affirmative professional discourse, another one of us recalls the impact of their sensory assessment report, which described the contributor as having 'vast issues with sensory overload'. This recognised and accepted their ways of achieving sensory regulation, including engagement in certain sexual behaviours, and so validated the individual and their choices. The language of the assessment not only provided them with:

> ... practical exercises that eventually led me back to a place of wellness both in my body and in my brain, [and] it also provided me with validation about why I had done the things that I did and why I continued to crave them.

This supported the practices that helped and affirmed their knowledge of how to cope with sensory experience. This is similar to the first story about the impact of epistemological travelling from a pathologising epistemological community to an empowering one. Here, the alternative language from affirmative professional communities made it possible to redefine themselves and their sensory experiences; from being perceived as pathological and possibly irrational 'problem behaviours' to rational responses to sensory needs. They said:

> Furthermore, the explanation for the behaviours that I had engaged in to meet my sensory needs throughout my life that had been labelled in turn, 'sick', 'perverse', and 'strange' was suddenly clear to me. The kink that I had engaged in had always been a source of great joy and pleasure but also conflict, as certain partners of mine have not shared them, and because of the opportunities for bullying and blackmail available to those who came to know of what I did. Although my kink did not manifest as a direct part of my autistic experience I too, like the other authors, have been caught between the attitudes and beliefs of a niche community, the BDSM community, and the psychiatric interpretations of my behaviour that were reflected in wider social attitudes about kinky sex and life style choices.

During analysis, this contributor specified:

> I did not reimagine the sensory report in an emancipatory way; rather the whole process was emancipatory and provided a positive and alternative narrative to the one I had been stuck with up until that point which was based upon an ICD-10 diagnosis of high-functioning autism.

In this way, behaviours that are diagnostically considered to be restricted and repetitive, often either described as non-functional (Ghanizadeh, 2010) or purposeless (Subki et al., 2017), are recognised as constituting valid sensory management strategies (Cunningham & Schreibman, 2008). In terms of the stranger, this implies the usefulness of collaboration between the stranger and non-stranger 'allies': while the stranger has the *potential* to be 'freer practically and theoretically; (surveying) conditions with less prejudice' (Simmel & Wolff, 1950, p. 402), this freedom may require an interactive emancipatory process, an engagement with discourses that open other possibilities to deficit readings. Previously, we saw how this was initially accessed via online autistic communities; here it is through interaction with a professional who recognises autistic experience as valid. This recognition brings far-reaching positive consequences:

> Finding out about why I did the things that I did not only gave me a more positive identification with it, I also understood *why* for the first time ever in my life. This enabled me to find positive alternatives that created less conflict within my intimate relationships and family dynamics.

This understanding of their own actions thus allowed for a 'self-reading' outside that of deficit. Similarly, the former contributor moved towards negotiating 'the closest approximation to my experiences and, indeed, the "entity" of autism', beyond language:

> I do what feels good, with enough confidence to know that I know what I need, far better than doctors, diagnostic manuals, and social stories. I've got myself here [...] via a sensory story that very few people could comprehend. I am beyond the definition of most people, even beyond myself using language, despite instinctively feeling like I'm figuring out the edges.

A sensory experience that seems to reflect 'a 'reality outside language' may accordingly also be the solution to the very difficulty it creates: 'kink, bondage, submission all became ways of solving this sensory conundrum, using a language that went beyond explaining'. Here, the act of submission became a linguistic 'replacement', a way of sharing sensory experience with others, in certain contexts, that went beyond the need for language and instead directly accessed the body's sensory ability to respond upon command.

> For me the submissive act becomes a way to share a sensory experience that is usually locked up within myself. It becomes a clumsy way for another to access what I have. They are not able to speak my language or have my cultural understanding but can clumsily interpret and direct what I experience through their control of my body and senses. It allows me to unlock the sensory world in which I exist and allow others firstly in and [it also allows me] to 'forget' about some of the sensory input that I always have to process through the act of sensory deprivation committed by another on me. The important thing is this latter part, I don't have to think about it, it takes no effort. It is a relief.

To get to this place took 'the loss of my rational mind and desire to live, to understand these things, because figuring them out was part of figuring my way back out of the darkness, into light and meaning again'. Meaning was created from sensory, not rational, language, thereby using nonverbal sensory experiences as a primary route back to wellness and self-acceptance.

> This speaks to me of a language that is specific to the time, space and people involved in this shared experience/creative linguistic process and reflects the unique way in which autistic people take sensory and environmental data that is very specific, instead of relying on past experience and mental knowledge schemas as NT people do.

This last quote illustrates the stranger's possibility of moving away from what may generally be accepted as truth, one way in which the stranger 'is not tied down in his action by habit, piety, and precedent' (Simmel & Wolff, 1950, p. 405). The stranger may be free – at a cost – to hold an alternative perspective and experience that may illuminate the commonly taken-for-granted. The stranger, here, forms new ways of nonverbal communication in order to make these alternative perspectives comprehensible and legitimate in different worlds, such as autistic or professional communities. As we have seen, this may be done through redefining and reimagining concepts and experiences, but also through epistemological alliances between epistemological communities.

Reflection on process – implications for methodology

Here we reflect upon some of the challenges and possibilities we encountered during our writing process. While our process did not necessarily set our work apart from neurotypical methods, we feel that there is a difference in the *sensibility* of this process. This is based on acknowledging and addressing needs from a shared assumption not of abilities and disabilities, and of different ways of functioning and the impact of different previous experiences in the group dynamic.

Most of us in the writing collective belong, or we have belonged to various degrees at differing moments in our lives, in BDSM spaces. Through this, some of us experienced a particular form of a shared senso-reality, an 'alternative social flow' during the writing process. This manifested itself in various ways. Most superficially it led to our initial interest in exploring BDSM experience in this chapter. However, more profoundly, our various positions in the BDSM world (as submissives or dominants, for example) started playing out during the writing process, and during both the initial one-day meeting and the residential retreats some of us felt a powerful sense of connection with each other and with our shared memories of BDSM scenes in which we had participated. This shared social flow *feels* like a sensory experience, as one of us puts it 'it is as though the air between us thickens and conducts our thoughts, emotions, and needs without passing via language or cognition', thereby functioning as nonverbal communication within the group. While some of us experienced this through our shared BDSM practice it could also conceivably be experienced between individuals sharing any intense interest, for example at an event during an autistic conference where autistic people get together to enjoy collective stimming and play with LED sparkly toys. To return to the notion of 'travelling otherwise' through sensory experience, this intense collective social flow brings with it the sense of travelling with others who co-create meaning not through language but through sensory experience, similar to the experience of the contributor discussing their kink experience in their earlier sensory story.

However, while this can be a powerful tool in a group writing process, it may also be difficult to manage, a sort of 'collective experience of monotropism' (see Murray, Lesser, & Lawson, 2005 for discussion of monotropism) that is so exciting and all-consuming, so deeply intimate and 'completing' that it becomes difficult to leave *as a group*. This intense immersion in an act, a moment, or a process brings many advantages and can contribute to autistic peoples' quality of life in various ways (McDonnell & Milton, 2014). However, when shared by some members of a group, but not all, in this way it can also be exclusionary, causing the shared nonverbal communication used by contributors *in the flow* to become unavailable to those *outside of the flow*.

This sense of exclusion was present at various stages for various members of our group. At an early point in our co-writing, two contributors were having difficulties in their own lives, and were therefore unable to participate to a great extent in the writing process. This meant that the two other members (who are also friends/collaborators outside the context of this writing group) spent considerable time together working on this project. This had various consequences. Our iterative process became one in which the two people who were initially less involved in writing up sometimes felt a stranger-like distance. As one of our contributors commented on reading an earlier draft,

> we are of the group and [share?] the perspective of the authors and data but remain outside of the intensive writing process you both undertook and so retain a certain objectivity and different perspective [insider/outsider] to the methods you describe here.

Thus, as with the stranger, being a 'full-fledged member [of a group] involves both being outside it and confronting it'. This required the two authors who spent considerable time together to revisit their analysis, to question their assumptions, and to reopen what they had thought of as finished. In this way, the theoretical notion of the stranger took form within our writing process: given the level of epistemological intimacy, and time, spent together by the two contributors, the other authors sometimes felt that they assumed parts of the role of Simmel's stranger, being of the group of authors and data creators, and yet remaining spatially, temporally, and sometimes emotionally and conceptually outside of the intensive writing process. The distance that had to some extent been created was therefore sometimes personally and relationally difficult, but also potentially powerful epistemologically.

Writing as a neurodivergent group required considerable flexibility. The member of the group who has the most difficulty with flexibility, who feels a 'rigid need for tight initial planning', is the only contributor who is not identified as autistic. However, within this group, that rigidity was also seen as something that needed to be taken account of, part of an 'increased awareness of the needs of others than might be expected to work in an NT group' as another contributor put it.

This awareness and negotiation of individual particularities played out on various fronts. During the work process we had several Skype meetings. In these, we each had differing needs (for example, one of us needed to 'pace' to think, another found the moving image difficult to process; one needed visuals, another found them overwhelming at moments). This required ongoing development of ways of communication and working together: for example, the division of the group into smaller entities during meetings (for example one-to-one-meetings) or a method of virtual 'co-writing' where two of the authors met simultaneously on Skype and worked together with the text in a shared google docs.

Many neurodivergent people struggle with their mental health (Griffiths et al., 2019), and this was also the case for members in this group. This required that different members of the group picked up the reins so as not to 'disqualify' anyone from participating, but this had to be approached pragmatically. This led to a level of mutual care that is perhaps uncommon in academic writing, but which in an autistic co-writing experience may be part of the process. It also underscores that perhaps a 'standard' external academic process (of, for example, a relatively tight and inflexible deadline schedule) may not always sit easily with a neurodivergent group of contributors.

Concluding reflections

From our writing process, new possibilities and challenges arose, with consequences for both research method and theorising. Starting with a common interest in exploring our sensory experiences in relation to sensory normate worlds and thinking, we found ourselves in somewhat new terrains, lacking both in clear methods and applicable theories, resulting in a pragmatic development of both methods and theoretical concepts 'as we went'.

Conceptualising The Stranger as an epistemic asset has been useful to us on these two counts. Methodologically it has allowed us to take some distance from, and to observe the effects of, the ways in which 'strangerhood' may play out in the writing process. This came with the sometimes painful awareness of the difficulties and emotional challenges present between members of our writing collective, whereby some of us, for a variety of reasons, became 'strangers' within our group. While this brought challenges, it also brought into play the usefulness of the stranger position, requiring that two authors who spent considerable time together revisited assumptions they had made from working 'at the centre' of the process.

Theoretically, too, we felt we advanced through our use of the stranger as a framework, which enabled – indeed required – distance from a deficit view of autism and in so doing permitted the emergence of two 'conditions' that may be necessary – or at least useful – in a process of autistic reconstruction. In a previous paper (Bertilsdotter Rosqvist & Jackson-Perry, 2020) two of the contributors explored the difficulties for individuals in describing their own (sexual) experience outside existing pathologising discourse. They asked: '(W)ithin the context of a body of autism literature which is generally deficit-driven, is it then possible for autistic people to imagine themselves and their intimate experience other than through deficit?' We noted here how one of our contributors was able, through their autistic community, to approach their own experience from a position of subjectivity, to move from this position to an understanding of that experience, rather than defining the experience through pre-existing, deficit-based theory. Echoing Belek (2019, p. 39) when he notes that the 'goal and challenge of many of my autistic interlocutors is to find ways to articulate their sensations', one of us noted the importance of a personal rethinking process; 'the process of owning the words that describe my own experiences', which when successful allowed 'a more complete and meaningful experience to emerge'. This rethinking process, which included a process of talking back to sensory normate conceptualisations, was enabled by access to – and identification with – autistic-created knowledges via both online and 'in real life' autistic communities, which may therefore constitute a first 'building block' of autistic identification.

Second, we noted the important role that may be played by professional 'allies'. The willingness of professionals to question and overcome deficit-based approaches, rethinking their theoretical knowledge in the light of their clients' autistic subjectivity, can be a powerful tool. Rather than the imposition of deficit, this brings an acknowledgement of difficulty while recognising the validity of autistic ways of working with those difficulties. In our contributor's example this meant considering kink, bondage, and submission not as further examples of their deficit, but as meaningful tools with which they knew, and navigated, their sensory needs and desires, an approach that brought affirmation to the autistic client and increased understanding to the professional.

In both these examples, a highly personal process of identity construction becomes part of an epistemological, and indeed political, shift. As strangers, contributors played the 'go-between' between worlds of autistic subjectivity, autistic community, and professional knowledge. Within research, we are increasingly seeing echoes of this type of endeavour of centring autistic subjectivity, a 'talking back' that positions autism as a place from which knowledge can be produced. However, we have seen here, even within a small group with shared ambitions and neurostati in common, that we have had to recognise the conditionality of sharing, of mutually exploring and discussing sensory experience. This may well be a major challenge that presents itself to neurodiversity studies as they move forward: 'talking back' is one thing, readily theorised through feminist, bisexual, and queer theories. Being heard, and then understood – on our own terms – is something quite different.

Notes

1 Reflecting the period in which these authors were writing, the stranger was written of as 'a man', using only masculine pronouns. We have retained original quotes here, knowing of course that the stranger is not of any particular sex or gender.
2 BDSM refers to bondage/discipline, domination/submission, sadomasochism.

References

Belek, B. (2019). Articulating sensory sensitivity: From bodies with autism to autistic bodies. *Medical Anthropology*, *38*(1), 30–43.
Bertilsdotter Rosqvist, H., & Jackson-Perry, D. (2020). Not doing it properly? (Re)producing and resisting knowledge through narratives of autistic sexualities. *Sexuality and Disability*. https://doi.org/10.1007/s11195-020-09624-5
Bogdashina, O. (2016). *Sensory perceptual issues in autism and asperger syndrome: Different sensory experiences-different perceptual worlds*. London: Jessica Kingsley.
Butler, J. (1990). *Gender trouble: Feminism and the subversion of identity*. New York: Routledge.
Chown, N., Robinson, J., Beardon, L., Downing, J., Hughes, L., Leatherland, J., Fox, K., Hickman, L., & MacGregor, D. (2017). Improving research about us, with us: a draft framework for inclusive autism research. *Disability & Society*, *32*, 720–734.
Clare, E. (2017). *Brilliant imperfection: Grappling with cure*. Durham, NC: Duke University Press.
Cunningham, A. B., & Schreibman, L. (2008). Stereotypy in autism: The importance of function. *Research in Autism Spectrum Disorders*, *2*(3), 469–479.
Davidson, J., & M. Orsini. (2013). Critical autism studies: Notes on an emerging field. In J. Davidson & M. Orsini (Eds.), *Worlds of autism: Across the spectrum of neurological difference* (pp. 1–28). Minneapolis, MN: University of Minnesota Press.
Dinishak, J. (2016). The deficit view and its critics. *Disability Studies Quarterly*, *36*(4).
Garland-Thomson, R. (2005). Feminist disability studies. *Signs: Journal of Women in Culture and Society*, *30*(2), 1557–1587.
Ghanizadeh, A. (2010). Clinical approach to motor stereotypies in autistic children. *Iranian Journal of Pediatrics*, *20*(2), 149–159.
Griffiths, S., Allison, C., Kenny, R., Holt, R., Smith, P., & Baron-Cohen, S. (2019). The Vulnerability Experiences Quotient (VEQ): A study of vulnerability, mental health and life satisfaction in autistic adults. *Autism Research*, 1516–1528.
Kafer, A. (2003). *Feminist, Queer, Crip*. Bloomington, IN: Indiana University Press.
McDonnell, A., & Milton, D. (2014). Going with the flow: Reconsidering 'repetitive behaviour' through the concept of 'flow states'. In G. Jones & E. Hurley (Eds.), *Good autism practice: Autism, happiness and wellbeing* (pp. 38–47). Birmingham, UK: BILD.

McRuer, R., & Johnson, M. (2014). Proliferating cripistemologies: A virtual roundtable. *Journal of Literary & Cultural Disability Studies*, *8*(2), 149–170.

Milton, D. (2014). Autistic expertise: A critical reflection on the production of knowledge in autism studies. *Autism*, *18*(7), 794–802.

Murray, S. (2012). *Autism*. New York: Routledge.

Murray, D., Lesser, M., & Lawson, W. (2005). Attention, monotropism and the diagnostic criteria for autism. *Autism*, *9*(2), 139–156.

Nadesan, M. H. (2005). *Constructing autism: Unravelling the 'truth' and understanding the social*. New York: Routledge.

Nelson, L. H. (1993). Epistemological communities. In L. Alcoff & E. Potter (Eds.), *Feminist epistemologies*. New York: Routledge.

Oikkonen, V. (2013). Competing truths: Epistemic authority in popular science books on human sexuality. *European Journal of English Studies*, *17*(3), 283–294.

Patsavas, A. (2014). Recovering a cripistemology of pain: Leaky bodies, connective tissue, and feeling discourse. *Journal of Literary & Cultural Disability Studies*, *8*(2), 203–218.

Rhys-Taylor, A. (2013). Developing a sociology of the senses (book review). *The Senses and Society*, *8*(3), 362–364.

Schütz, A. (1944). The stranger: An essay in social psychology. *American Journal of Sociology*, *49*(6), 499–507.

Serlin, D. (2017). Science and the senses: Deviation. *Cultural Anthropology*. URL: (Retrieved December 2019) https://culanth.org/fieldsights/1070-science-and-the-senses-deviation

Simmel, G., & Wolff, K. (1950). *The sociology of Georg Simmel*. London: Collier-Macmillan.

Subki, A. H., Alsallum, M. S., Alnefaie, M. N., Alkahtani, A. M., Almagamsi, S. A., Alshehri, Z. S., Kinsara, R. A., & Jan, M. M. (2017). Pediatric motor stereotypies: An updated review. *Journal of Pediatric Neurology*, *15*(4), 151–156.

Vannini, P., Waskul, D., & Gottschalk, S. (2013). *The senses in self, society, and culture: A sociology of the senses*. London: Routledge.

Ward, J., Hoadley, C., Hughes, J. E. A., Smith, P., Allison, C., Baron-Cohen, S., & Simner, J. (2017). Atypical sensory sensitivity as a shared feature between synaesthesia and autism. *Scientific Reports*, *7*(41155).

Yergeau, M. (2018). *Authoring autism*. Durham, NC: Duke University Press.

Part IV

Neurodiversity at work

Chapter 9

Practical scholarship
Optimising beneficial research collaborations between autistic scholars, professional services staff, and 'typical academics' in UK universities

Nicola Martin

Practical collaborative scholarship

My research, focused on improving disability equality in post-compulsory education, coalesces around the idea that practical collaborative scholarship is the way
forward. An insider perspective is key to this area of enquiry. Those tasked with
delivering services designed to make higher education more inclusive have essential insights to share, alongside disabled staff and students. Collaboration between
interested parties in different roles colours a vivid picture. Practical scholarship
requires that all players are committed to ensuring that research outcomes are
translated into recommendations for positive change. As someone who now
passes as a 'typical academic' by virtue of my job title, I recognise my privilege.
My own neurodivergence and atypical career trajectory, including many years
managing professional services, inform my position.

In this chapter, focusing specifically on the contribution of autistic scholars and
researchers based in Professional Services (PS) teams, I trouble structures and
attitudes that make collaborative insider informed research difficult. I go on to
propose possible solutions. Relevant definitions (including the term 'academic')
are critically unpacked in the following section, which challenges the idea of a
shared understanding of some key terms in common usage in academia.

Status and problematic definitions: ally, academic, ableism, activist, merit

Easy as it is to adopt the rhetoric of allyship and activism, self-identifying as an
ally/activist is somewhat presumptuous. My job description unambiguously says
'academic'. An exclusionary undercurrent is created by definitions of 'academic'
that disenfranchise scholars not firmly attached to university academic roles.

The *Oxford English Dictionary* (Oxford, 2019) definitions for the terms activist, ally, and academic (in noun form) require some unpacking. An activist is

defined as 'A person who campaigns to bring about social or political change'. Ally refers to 'A state of formally co-operating with another'. Academic is used to name 'A teacher or scholar in a university or other institution of higher education'. The implication that 'an academic' must be formally associated with the HE establishment, presumably as a student, lecturer, or researcher, is contentious. While acknowledging the high quality of much autistic scholarship, this paper raises concerns about the status of autistic researchers in relation to the academy. Assumptions about academic-professional services' role divisions are also problematised for their potential to stifle collaborations because of the convention that only (narrowly defined) academics engage in research.

Bourdieu (1998) uses the term 'title of nobility' in theorising Cultural and Social Capital. He asserts that benefits and privileges are bestowed upon individuals with certain titles in the form of status and associated opportunity (Bourdieu & Farage, 1994). Researchers have utilised Bourdieu's ideas of Capital when considering aspects of student experience and education policy but have not turned the same spotlight towards the questions under discussion here (e.g. Maton, 2005). Interest in addressing structural barriers to research collaborations may be limited by funding or by unhelpful perceived status differences between 'academic' and 'academic related' staff. 'Professor' is a high-status title linked with a particular set of expectations around research activity. My fancy title opens doors. Despite a certain degree of status associated with leadership roles, as head of a PS department I sometimes found myself having to bang assertively on those doors.

Status is defined in the *Cambridge English Dictionary* (Cambridge, 2019) as 'an accepted or official position, especially in a social group: the amount of respect, admiration, or importance given to a person'. Status, power, and privilege are first cousins. Lack of status can create obstacles that powerful privileged people can choose to demolish. In relation to autistic scholarship, a more pertinent indicator of status may be the quality and influence of scholarly works rather than the professional roles of their authors. Merit and meritocracy are useful concepts in relation to the idea of status. The *Cambridge English Dictionary* (Cambridge, 2019) defines merit as: 'The quality of being good and deserving praise'. 'A meritocracy is a society or social system in which people get status or rewards because of what they achieve, rather than because of their wealth or social status' (Collins, 2019). Divorcing considerations about the status of a piece of work from the job titles of the author, and concentrating on its academic merit, could partially address the disenfranchisement of scholars unlikely to be employed towards the top of the academic tree.

Academic freedoms enabling me to engage in research congruent with my values are associated with privileges, including my role and, to an extent, my age. At 60, with most of my career behind me, CV-building is not my main motivation, so I can make choices about where I publish and with whom I collaborate. Usefulness potential underpins these decisions. Privilege is multi-faceted and I recognise that the burden of student debt associated with higher education is also something that did not concern my generation (Murphy, Scott-Clayton, & Wyness, 2019).

A disparity in access to employer-funded higher degree study between lecturers and PS staff has been reported via The National Association of Disability Practitioners (NADP) JISCmail list, illustrating further the relative advantage of academic titles and established roles.

In the UK, HE compares institutions using a range of measures, including the TEF (Teaching Excellence Framework).[1] Student experience and the learning environment are key TEF indicators of quality, which include an equalities dimension relevant to neurodivergent learners. Disenfranchising PS staff from research designed to improve student experience makes for an incomplete picture, even when student voice is included. Results of TEF inspections are in the public domain, and a factor in student choice, so could provide leverage in the argument for collaborative research.

In Foucauldian terms, an approach that impacts negatively on those without perceived status and the accompanying associated power and privilege would be categorised as 'othering' (Foucault, 1982). Othering is a pejorative term that detrimentally constructs 'them' (autistic people and PS colleagues in this case), in relation to 'us' (academic university staff who are expected to carry out research). Sometimes the 'we'-and-'they'–binary gets muddled up. The existence of autistic lecturers and researchers and disability advisors with doctorates and publications may surprise anyone who did not realise that the academic club was open to diversity.

Obstacles that autistic people, however well qualified, encounter in accessing and sustaining paid employment make the status of employee hard to attain and sustain. Although comprehensive data is unavailable, evidence from The Participatory Autism Research Collective (PARC) indicates that autistic doctoral students rarely progress to research and lecturing contracts (Gartsu & Stefani, 2019; Harmuth et al., 2018). If academic is defined in terms of a formalised relationship with the academy as either student or staff, those without such an association are effectively disenfranchised.

Grudgingly opening the door a little bit is not enough. Campbell (2009) and others use the term 'ableism' to denote attitudes and societal constructs that can impact adversely upon disabled people. Loja, Costa, Hughes, & Menezes (2013) equate ableism with 'The invalidation of impaired bodies and the constant struggle to establish credibility' (p. 193). Universal Design (UD) principles are built on the avoidance of ableist assumptions and the expectation that planning will take diversity into account (Milton, Martin, & Melham, 2016). The Equality Act (2010) covers universities and is congruent with UD thinking. Alongside the TEF, the requirement of the Equality Act to listen to stakeholders as part of planning for diversity reinforces the case for inclusive research. Expectations under the Act include Equality Impact Assessment of policy, practices, and procedures that are informed by those likely to be affected. Impact assessment is an avenue for insider perspective research.

Inclusive practice extends beyond increasing the diversity of the workforce and student body. Neurotypical privilege has been identified by neurodivergent researchers who point to the failure of the academy to engage in practices that accommodate a widening range of approaches and thinking styles (Bertilsdotter Rosqvist et al., 2019). Bashing square pegs into round holes because 'that is just the way we do things around here' is problematic and a waste of talent (Harmuth et al., 2018). Collaborating with autistic and other neurodivergent researchers involves avoiding the ableist practice of expecting everyone just to fit in, put up, and get on, rather than developing a supportive collegiate working environment that values everybody's talents and creates conditions conducive to their best contribution (Gartsu & Stefani, 2019).

Autistic and neurodivergent employees in various contexts have discussed situations in which, in order to fit in, they have felt it necessary to mask characteristics that may be socially constructed as indicative of their otherness (Patton, 2019). Masking could take the form of pretending to understand and therefore not asking for help for fear of exposure. The general agreement is that masking can at best cause individuals unnecessary pressure and at worst can impact extremely negatively on wellbeing and mental health (Milton & Sims, 2016). Although my interest faces towards social justice concerns, some equitable employment practice research focuses on 'the business case' for working practices conducive to wellbeing (Martin, Milton, Sims, Dawkins, & Baron-Cohen, 2017). Logic dictates that employees who feel alienated and uncomfortable are less likely to perform effectively at work. A lot of energy goes into masking. In an inclusive workplace this would be unnecessary.

Carving out the contours of practical scholarship as a possible way of working as an ethical and reflexive academic involves engaging with contentious labels such as ally and activist. Scholarship feels like a less controversial term than activism when applied to work undertaken from my privileged position. Engaging in activism is altogether more risky for those without my salary safety net. Consequently, I do not claim the title of activist, which arguably should not be open to self-nomination anyway. 'Practical scholarship' better describes my research, which is designed for usefulness rather than shelf filling. Examples of approaches to practical scholarship are discussed in the next section.

PARC as a structure to enhance opportunities for autistic researchers

Under the umbrella of the London South Bank University (LSBU) Centre for Social Justice and Global Responsibility (CSJGS) I lead the 'Critical Autism and Disability Studies (CADS)' research group. Structurally PARC was originally part of CADS, which includes disabled and non-disabled academics and PS staff. CADS works on the principle of 'nothing about us without us' (Charlton, 1998) and overtly leans towards the social model concerns of eradicating societal barriers in the name of social justice (Oliver, 2009). My role alongside Dr Damian Milton in the development of the Participatory Autism Research Collective (PARC) (https://participatoryautismresearch.wordpress.com/about) is essentially supportive. Dr Milton and other autistic scholars were and are the main drivers. PARC started at LSBU in 2015 as a vehicle for autistic scholars to work together. The collective now operates in various locations across the UK and is extending its reach into Europe and America. Occasionally I find myself representing PARC informally when attending an event for another purpose, usually outside the UK. This is pragmatic rather than ideal as, although neurodivergent, I am not autistic. Study trips to America and Europe are further indicators of my privilege.

PARC has delivered annual critical autism studies conferences at LSBU since 2017. The website provides conference details as well as examples of numerous seminars and researcher development activities. Autistic presenters are in the majority and contributions from neurotypical researchers are only included if selected by autistic PARC members. All events are free because many people wishing to participate do not have the backup of an employer willing to meet attendance expenses. Provision of facilities, such as a quiet room, is built into the planning to respect the requirements of attendees who may experience sensory overload (Milton, Ridout, Kourti, Loomes, & Martin, 2019).

Universities support PARC by hosting events without charging for rooms and the collective could not function effectively otherwise. The danger of overpromising and under-delivering is real when operating on a shoestring. Conferences are organised along good autism practice principles such as avoiding sensory overload, but occasionally overcrowding occurs and rooms get too hot. Because PARC is not paying for conference space, choice is limited, leading to logistical nightmares such as the quiet room being located on a different floor from the conference space.

Catering costs are a thorny practical issue. Autistic delegates find the idea of having to use university caterers illogical when they could easily create a 'bring and share' feast from the local supermarket. Various dietary requirements of participants could be easily catered for if individuals each contributed something suitable for their own needs. This would minimise the potential for anxiety about uncertainty that some autistic people experience around food. Sometimes, to save money, conferences build in longer breaks and supply lists of local eateries. For delegates not confident in unfamiliar surroundings this would be impractical.

LSBU opens up the events to students interested in autism and their evaluations evidence that PARC adds value to the curriculum. Similar arrangements apply in other universities. PARC was not conceived as a resource for non-autistic students and their participation has to be handled with care. In practice, students are respectful and informed by social model principles and no difficulties have arisen.

Evidence of adverse impacts on self-esteem through perceived exclusion by peers reminds us that social aspects of college and university can be toxic, or (with a few tweaks) far more enabling (Chown, Baker-Rogers, Hughes, Cossburn, & Byrne, 2018; Milton & Sims, 2016). Autistic people trying to manage the alienating environment of their university or workplace frequently describe the feeling of 'othering' (Foucault, 1982; Richards, 2008). One of the advantages of PARC is that it provides a safe space in which masking is not necessary. Ideally, camouflaging in the workplace would always be unnecessary and the aim of much of my research is to provide a basis on which to move forward with this aspiration.

Members of PARC largely conform to an inclusive and meritocratic definition of academic, which acknowledges the scholarly nature of research often undertaken despite myriad ableist barriers, including lack of appropriate renumeration or effective reasonable adjustments in the workplace. Academic outputs from PARC members, in the form of refereed journal articles, books, chapters, conference presentations and so on, speak for themselves as evidence of impactful, high status scholarship (see e.g. Chown et al., 2018; Lawson, 2017; Milton et al., 2019).

The list of references just quoted illustrates the productivity of PARC members, even though these examples represent a tiny fraction of the output. Unfortunately, eligibility rules for inclusion in Research Excellence Framework (REF)[2] submissions disenfranchise most of the authors just mentioned. Rather than being a quality issue, the problem is lack of employment contracts acceptable to the REF. Having work included in the REF is a university esteem indicator but clearly there are blocks in the road that have nothing to do with scholarly merit. Co-authoring enables profile-raising for researchers not specifically named in REF submissions. An alternative interpretation is that co-authoring benefits disproportionately those already recognised by the academy.

Authenticity is an underpinning PARC value that involves striving to find ways of moving beyond tokenistic inclusion of people who communicate in ways that are not easily accommodated within the academy (Milton et al., 2019). Indeed, communication of personal aspirations without voice was the theme of an excellent doctoral thesis by Brett (2016), which I have had the privilege to supervise recently, and which will be the topic of a forthcoming PARC seminar.

PARC, although a major contributor, is not the only outlet for research conducted by autistic people and the phenomenon is not recent. The ASPECT survey (Beardon & Edmonds, 2007), for example, was controlled by autistic people. Over 200 ASPECT participants shared their perspectives about quality-of-life indicators. Resulting rich data covered themes such as housing, education, and criminal justice. ASPECT was unfunded and is a salutary lesson in how much more autistic researchers could do with a little financial backing.

NADP: a vehicle for collaborative research

Practical scholarship characterises my 20-year engagement as a board member of the National Association of Disability Practitioners (NADP) (https://nadp-uk. org). Originally, NADP was set up as a professional association to help staff to work effectively with disabled students (Wilson & Martin, 2017). With over 1,500 members, mainly from PS roles but including UK and international academics, NADP has become the go-to network for those committed to disability equality in post-compulsory education. Although research teams that include PS staff as well as academics are still a rarity (Chown et al., 2018), NADP acts as a catalyst for such collaborations. My position enables me to progress this agenda. As with PARC, volunteers are the lifeblood of the organisation. NADP, unlike PARC, charges a membership fee. This funds administrative support, which is a significant enabler not available to PARC at this stage.

Professional development is central to NADP and particularly important for PS staff who often have limited access to other outlets. Activities include: training events, networking via JISCmail, and highly regarded international conferences in which the voices of disabled staff and students are heard. Inclusive practice is built into process as well as content for NADP events. Social model thinking underpins NADP and the practical scholarship that emanates from the membership coalesces around the idea of removing barriers experienced by disabled people in college and university. My personal contribution includes numerous conference presentations and training activities designed to build the research skills and confidence of PS colleagues.

Researching with diverse teams in which everyone brings their own ontological perspective to the party can prove challenging. Doing so enables a wider view than would be likely if a group of people all sat around the table agreeing with each other. While it is easy to paint a picture in which the challenge is to the academy for not allowing PS staff to carry out research and failing to pay autistic researchers, in my experience the reality is somewhat more nuanced. Listening and talking to each other with open minds across disciplines and role divides can open the door to new understandings. Discussing potentially contradictory assumptions as well as areas of common ground is an important aspect of working collaboratively. Understanding the context of the shifting landscape of UK higher education, in which a decade of equalities legislation has influenced the development of inclusive practice (Draffan, James, & Martin, 2017; Wilson & Martin, 2017), is essential in order to implement and evaluate change effectively. My ally allegiances face in various directions and coalesce around the idea of encouraging collaboration.

Academics in critical disability studies tend to frame discussions around the social model and PS staff talk the language of Universal Design for Learning (UDL) (Draffan et al., 2017). Both groups appear to be positioned in the eradicating barriers camp but the marriage is not without tensions. Some disability practitioners may argue that they have to work within a system requiring a diagnostic label to support disabled students effectively, while others challenge the orthodoxy of disabled naming as a gateway to services and therefore a necessary evil within HE (Martin, 2008). PARC members are out and proud about their autism, but some autistic people are not as comfortable with the label and consequently less willing to contribute to research. Other autistic scholars do not disclose because of the risk to their careers if they do so. Many autistic voices are therefore effectively silenced.[3] Services for autistic students and staff who are not out and proud are potentially behind a wall if they are dependent on a paradoxically medical-model diagnostic process, in a notionally inclusive context (Hastwell, Harding, Martin, & Baron-Cohen, 2013).

I am privileged to have been part of several effective research collaborations between academic and PS staff and students. These all operated without unhelpful undercurrents around academic status. At Sheffield Hallam University (SHU) in the early 2000s, I was on a permanent academic contract as a principal lecturer (PL) while heading services for disabled students. Unusually, the team included hybrid staff who researched and lectured in critical disability studies as well as supporting disabled students directly. We worked closely with academics in the SHU Autism Centre on insider-informed research, focused on disability equality concerns within the academy. Autism Centre academics continue to contribute to staff development around good autism practice in higher education at SHU (Chown et al., 2018). The helpful structure at SHU is something I inherited from Clive Owen (former Head of SHU Student Services) and so claim no credit for the vision and strategy from which it was created. Collaborations that began at SHU have endured.

Professor Simon Baron-Cohen led a project designed to improve the university experience of autistic Cambridge students by asking them for their expert advice and using the information to underpin change (Hastwell et al., 2013). Baron-Cohen sought funding for a researcher post, which he located within the Disability Resource Centre (DRC), and appointed a disability practitioner to the position. He was adamant that autistic students should also participate in the steering group but this proved to be easier said than done. The absence of a paid autistic researcher was a limitation of the Cambridge project (Hastwell et al., 2013).

Collaborative research enables productive conversations that might otherwise not happen. Academics investigating how autistic people experience university need insights into factors beyond the classroom that PS providers and end users can offer. Autistic researchers can advise on methodology and ensure that the right questions are asked in the right way.

Vision and strategy are required in order to facilitate cooperative productive working, which overcomes barriers between roles. Structural splitting can stop disability practitioners from being researchers. An insider perspective may not be factored in adequately if it is not part of the research vision and strategy. Staff in PS roles, sometimes disparagingly referred to as 'non-academic', autistic scholars, and others need encouragement, support, and an effective infrastructure in order to develop their confidence as researchers (Martin et al., 2017).

Maslow and Lewis (1987) originally identified belonging as necessary for progress to self-actualisation in the 1950s. Belonging is built into the foundations of universal design and social inclusion. Belonging implies community. While student experience is more likely to be the focus of research, we must not forget that disabled staff are part of the community too and often face similar barriers (Martin et al., 2017; Milton & Sims, 2016). PS staff who find themselves justifying their own legitimacy as players within the research arena can also find themselves feeling disenfranchised within their workplace (Martin et al., 2017). An important part of my contribution is around community-building. Ultimately, for research to be of any practical use, those in a position to enact change need to find it, hence the birth of JIPFHE.

JIPFHE: making practical scholarship visible

The *Journal of Inclusive Practice in Further and Higher Education (JIPFHE)* is open access and peer reviewed and serves largely as a vehicle for 'practitioner research' (https://nadp-uk.org/resources/publications/published-journals/). It is the main outlet for much of the practical scholarship of NADP members and includes contributions from disabled students and other stakeholders. In my editorial role I am proactive about helping scholars to shape their contributions if necessary. The editorial guidelines emphasise the goal of usefulness, the requirement for accessibility, plain English, and insider perspective. The publication includes multiple examples of unfunded research undertaken by staff without 'researcher' in their title and unlikely to feature in the REF.

JIPFHE has not been tempted to publish work by academics seeking to understand the causes of autism. In keeping with PARC participants and other social modelists, the editors have no interest in medical-model perspectives around autism. Genetic origins are irrelevant to improving universities and the notion of finding 'a cure' feels rather insulting to autistic colleagues and students. It would be difficult to claim ally credentials alongside any sympathy with the idea of finding a way to eradicate autism in future generations. Exposure to PARC would allow those looking for a cure access to alternative perspectives.

While vanishingly few examples exist of scholars in critical disability/autism studies researching alongside PS colleagues, there are exceptions (e.g. Draffan et al., 2017; Hastwell et al., 2013). JIPFHE aims to bring disparate voices together, regardless of job titles and the status these convey. Universal Design is concerned with every aspect of university for every person, and *JIPFHE* shares the same ethos, so the editorial board welcome articles that look at the spaces beyond the classroom (e.g. Pritchard, 2017).

Practical scholarship?

Practical scholarship, which is informed by insider perspectives and conceived with the end goal of usefulness, can be helped along by structures such as those discussed here. While organisations like NADP and journals like *JIPFHE* can make small inroads into addressing silo working, which creates barriers to collaborative research between academics and PS staff, their influence is limited. 'Practitioner research' is a term that is sometimes used disparagingly in academia. *JIPFHE* is essentially a vehicle for practitioner research that is hardly on the radar in relation to the REF. Although PARC is making great strides in terms of recognising and facilitating the academic contributions of autistic scholars, the ableist barriers discussed here are still all too common. Despite being proactive about improving conditions for autistic researchers, CADS cannot claim to be a truly emancipatory research group (French & Swain, 1997). To claim emancipatory credentials, disabled researchers would need to be employed as principal investigators in order to have a greater degree of control of the research agenda. 'Researcher as parasite' is an expression that was coined by Stone and Priestley (1996, p. 699). While autistic researchers often do not enjoy parity of esteem with many other academics because of the nature of their employment contracts, parasitism is a concern. My leadership role has given me the power to make policy decisions about only engaging with autism research that includes paid autistic researchers, and to cast the net beyond a narrow group of collaborators from the pool of critical disability studies colleagues. Current large-scale CADS autism research includes employed autistic researchers, as will future projects.

PS colleagues are increasingly engaging in work-related research and postgraduate study and I am well positioned to help, having worked in disabled student support myself. My academic role has enabled me to develop an MA in education/autism and Education Doctorate (EdD) in social justice and inclusive education. Autistic students add something to the mix. Inclusive practice is embedded into the delivery as well as the curriculum, and the majority of the readings on the MA are written by autistic scholars. The EdD is a practice-based doctorate and provokes other conversations, which interest me very little, about its status in relation to a PhD. Losing sleep about whether practitioner research is REFable, or whether an EdD is the academic equivalent of a PhD, is not something that troubles me at all. I am too busy thinking about whether my contribution is of any practical use.

In the spirit of usefulness, I invented the acronym REAL which stands for: reliable, empathic, anticipatory, and logical. The REAL idea appears regularly in my research with autistic students but applies equally to academics who are not students. REAL underscores the idea that autistic people benefit from quite straightforward reasonable adjustments, which are also useful for everyone else (Martin, 2008).

Conclusion

Practical scholarship that places the question 'so what?' at the centre of the enquiry, has the potential to make a difference to people's lives. This will only happen if the stage beyond 'so what?' is enacted so that recommendations are translated into an action and evaluation cycle that ultimately underpins sustainable change. Inclusion and insider perspectives need to inform practical research in order for a sufficiently rounded picture to emerge. Barriers and enablers within this endeavour are chewed over in this chapter. My feeling is that it ought to be easier than it is to make practical scholarship of the type discussed here relatively commonplace. Structures such as PARC and NADP have the potential to generate a critical mass and a sense of solidarity between often-marginalised scholars. Rather than being systemic pillars of the university landscape these networks have grown organically from grass roots activism, and the contribution of volunteers merits recognition. Sustainability is inevitably an issue.

Autistic researchers need to be at the forefront of autism research and properly remunerated. I recognise my privilege in being in a position to do something about this, albeit on a small scale. Academics in critical disability studies, PS colleagues, and autistic staff and students researching together provide a wider lens with which to consider ways to improve universities for autistic people. I am motivated to encourage and facilitate such collaborations. As someone who has the advantages and status of a salaried research leadership position, I am well placed to be able to do something useful in this regard.

Acknowledgements

To all my fantastic autistic colleagues and to the people who make PARC a success through their voluntary efforts and indisputable brilliance.

Notes

1 www.thecompleteuniversityguide.co.uk/universities/choosing-a-university/teaching-excellence-framework-(tef)
2 The Research Excellence Framework is the UK system for assessing the quality of research in higher education institutions. It first took place in 2014. The next exercise will be conducted in 2021 (www.ref.ac.uk/about).
3 There are over 150 members of the Autistic Autism Researchers (secret) Facebook group. Although some members do disclose publicly, the mere fact that there is a need for this group speaks volumes about the stigma attached to autism.

References

Beardon, L., & Edmonds, G. (2007). *ASPECT consultancy report. A national report on the needs of adults with Asperger syndrome.* URL: (Retrieved July 2018) www.sheffield.ac.uk/polopoly_fs/1.34791!/file/ASPECT_Consultancy_report.pdf

Bertilsdotter Rosqvist, H., Kourti, M., Jackson-Perry, D., Brownlow, C., Fletcher, K., Bendelman, D., & O'Dell, L. (2019). Doing it differently: emancipatory autism studies within a neurodiverse academic space. *Disability & Society, 34*(7–8), 1082–1101.

Bourdieu, P. (1998). *The state nobility: Elite schools in the field of power.* Redwood City, CA: Stanford University Press.

Bourdieu, P., & Farage, S. (1994). Rethinking the state: Genesis and structure of the bureaucratic field. *Sociological theory, 12*(1), 1–18.

Brett, S. (2016). Future selves: listening carefully to the voice of a key stage 5 pupil in a special school. In D. Milton & N. Martin (Eds.), *Autism and intellectual disabilities in adults Vol. 1*, 5–61. Hove: Pavilion.

Cambridge (2019). *Cambridge English Dictionary.* Cambridge: CUP.

Campbell, F. (2009). *Contours of ableism: The production of disability and abledness.* Basingstoke: Palgrave Macmillan.

Charlton, J. I. (1998). *Nothing about us without us: Disability oppression and empowerment.* Berkeley, CA: University of California Press.

Chown, N., Baker-Rogers, J., Hughes, L., Cossburn, K. N. & Byrne, P. (2018). The 'High Achievers' project: an assessment of the support for students with autism attending UK universities. *Journal of Further and Higher Education, 42*(6), 837–854.

Collins (2019). *Collins English Dictionary*. New York: Collins.

Draffan, E. A., James, A., & Martin, N. (2017). Inclusive teaching and learning: What's next? *Journal of Inclusive Practice in Further and Higher Education, 9*(1), 23–34.

The Equality Act (2010). URL: (Retrieved August 2019) www.disabilityrightsuk.org/understanding-equality-act-information-disabled-students

Foucault, M. (1982). The subject and power. *Critical Enquiry, 8*(94), 777–795.

French, S., & Swain, J. (1997). Changing disability research: Participating and emancipatory research with disabled people. *Physiotherapy, 83*(1), 26–32.

Gartsu, G. N., & Stefani, M. (2019). *Tapping into people with autism spectrum disorder: Moving Towards an Inclusive and Neurodiverse Workspace*. (Unpublished master's thesis). KTH Royal Institute of Technology, Stockholm. URL: (Retrieved November 2019) http://kth.diva-portal.org/smash/get/diva2:1328726/FULLTEXT01.pdf

Harmuth, E., Silletta, E., Bailey, A., Adams, T., Beck, C., & Barbic, S. P. (2018). Barriers and facilitators to employment for adults with autism: A scoping review. *Annals of International Occupational Therapy, 1*(1), 31–40.

Hastwell, J., Harding, J., Martin, N., & Baron-Cohen, S. (2013). *Asperger syndrome student project, 2009–12: Final project report, June 2013*. URL: (Retrieved June 2018) www.disability.admin.cam.ac.uk/files/asprojectreport2013.pdf

Lawson, W. B. (2017). Issues of gender & sexuality in special needs children: Keeping Students with autism and learning disability safe at school. *Journal of Intellectual Disability-Diagnosis and Treatment, 5*(3), 85–89.

Loja, E., Costa, M. E., Hughes, B., & Menezes, I. (2013). Disability, embodiment and ableism: stories of resistance. *Disability & Society, 28*(2), 190–203.

Martin, N. (2008). REAL services to assist university students who have Asperger syndrome. *NADP Technical Briefing 10/08*. URL: (Retrieved July 2019) https://nadp-uk.org/wp-content/uploads/2015/04/REAL-Services-to-assist-students-who-have-Asperger-Syndrome.pdf

Martin, N., Milton, D., Sims, T., Dawkin, G., Baron-Cohen, S., & Mills, R. (2017). Does 'mentoring' offer effective support to autistic adults? A mixed methods pilot study. *Advances in Autism, 3*(4), 229–239.

Maslow, A., & Lewis, K.J. (1987). Maslow's hierarchy of needs. *Salenger Incorporated, 14*, 987.

Maton, K. (2005). A question of autonomy: Bourdieu's field approach and higher education policy. *Journal of education policy, 20*(6), 687–704.

Milton, D., Martin, M., & Melham, P. (2016). Beyond reasonable adjustment: Autistic-friendly spaces and Universal Design. In D. Milton & N. Martin (Eds.), *Autism and intellectual disabilities in adults, Volume 1* (pp. 81–86). Hove: Pavilion.

Milton, D., & Sims, T. (2016). How is a sense of well-being and belonging constructed in the accounts of autistic adults? *Disability and Society, 31*(4), 520–534.

Milton, D., Ridout, S., Kourti, M., Loomes, G., & Martin, N. (2019). A critical reflection on the development of the Participatory Autism Research Collective (PARC). *Tizard Learning Disability Review, 24*(2), 82–89.

Murphy, R., Scott-Clayton, J., & Wyness, G. (2019). The end of free college in England: Implications for enrolments, equity, and quality. *Economics of Education Review, 71*, 7–22.

Oxford. (2019). *Oxford English Dictionary*. Oxford: OUP.

Oliver, M. (2009). *Understanding disability, from theory to practice* (2nd ed.). Basingstoke: Palgrave Macmillan.

Patton, E. (2019). Autism, attributions and accommodations. *Personnel Review*, *48*(4), 915–934.

Pritchard, E. (2017). Body size and higher education: the experiences of an academic with dwarfism. *The Journal of Inclusive Practice in Further and Higher Education*, *8*(1), 5–13.

Richards, R. (2008). Writing the othered self: Auto ethnography and the problem of objectification in writing about disability and illness. *The Journal of Qualitative Health Research*, *18*(12), 1717–1728.

Stone, E., & Priestley, M. (1996). Parasites, pawns and partners; disability research and the role of non-disabled researchers. *British Journal of Sociology*, *47*(4), 699–716.

Wilson, L., & Martin, N. (2017). Disabled student support for England in 2017. How did we get here and where are we going? A brief history, commentary on current context and reflection on possible future directions. *Journal of Inclusive Practice in Further and Higher Education*, *9*(1), 6–22.

Chapter 10

Designing an autistic space for research

Exploring the impact of context, space, and sociality in autistic writing processes

Hanna Bertilsdotter Rosqvist, Linda Örulv, Serena Hasselblad, Dennis Hansson, Kirke Nilsson, and Hajo Seng

Introduction

The aim of this chapter is to explore neurodivergent writing processes, both from the perspective of separate neurodivergent individuals working more separately in *one´s own space* and as a neurodivergent collective, working together in a *shared autistic space* (Sinclair, 2010). The chapter is based on narratives referring to previous experiences of writing processes, and on metanarratives, reflecting upon the writing process at hand: the writing of this chapter. We will discuss the narratives from the perspectives of autistic autism theories – theories developed by autistic researchers about autism: principally Sinclair's social theory of autistic socialising in autistic spaces in combination with Seng's (2019) more cognitive theory of autistic thinking styles.

We found Sinclair's distinction between being autistic in one's own space or in a shared autistic space central during the process of working on this chapter, but we have also come to the insight in our process of the importance of meeting up in a shared *physical* space. By a shared (physical) autistic space in this chapter, therefore, we refer to a social-relational, as well as a physical, space where 'autistic people are in charge', where autistic people 'determine what our needs are', 'make the decisions about how to go about getting our needs met', and where we are being 'autistic together' (Sinclair, 2010). This shared autistic space is here contrasted not only with working in a neurotypical (NT) space, but also with working in other autistic spaces. On one hand, an autistic space defined by Sinclair as 'being autistic in one´s own space', on the other, a shared virtual autistic space (cf. Bertilsdotter Rosqvist, Brownlow, & O'Dell, 2013). Sinclair notes that the problems with 'being autistic in NT space and being autistic in one´s own space both involve being "the only one" – the only autistic person in the environment' (Sinclair, 2010). Those problems have emerged as central in our discussion throughout the project, as it was our experience that the shared physical autistic space helped the authors to overcome barriers and be more creative where the solo or virtual space was insufficient.

Seng's (2019) cognitive theory of autistic thinking focuses on an atypical connection between perception and processing. The theory stresses diversity and variability, offering alternatives to deficit-based models of neurodivergent cognition. Seng argues that autistic people's thinking separates the conceptualisation (language) part from a perceptual part that takes place on a sensory level. This is in contrast to non-autistic people where the perception and the concept are intertwined in such a way that it is difficult to distinguish them. Seng thus distinguishes between perceptual thinking and linguistic thinking (which, as he points out, is at least partly supported by recent neurobiological research). The autistic person, he argues, needs to put more effort into approaching linguistic thinking and creating words and concepts based on their perceptual experiences. The autistic perception of different concepts and words in the language then becomes more of a collage of parts than an experienced clear entity. An autistic person's exploration of the world is therefore more unconditional, less prejudiced, since the person is not strongly bound to the concept world and thus does not process sensory information according to predetermined concepts and framings to the same degree as neurotypical people. Autistic people approach the world as phenomenologists and proceed more from what the senses say than from learned interpretation patterns. This idea is a further development from the framework of predictive coding theory or 'Bayesian decision theory' (Pellicano & Burr, 2012). From this framework, the use or precision of prior knowledge is reduced among autistic people (Brodski-Guerniero et al., 2018; Haker, Schneebeli, & Stephan, 2016). However, in Seng's theorising it is not about deficits or impairments, but a less biased way of processing information.

In contrast to most ideas focusing on differences in cognitive styles between autistic and neurotypical people (for example Valla & Belmonte, 2013), Seng's theory also entails an empirically based typology of different autistic thinking styles, thus presenting a more diverse image of the autistic population. The different thinking styles depend on the type of perception that dominates for the individual. Despite the great heterogeneity of thinking styles, Seng argues that autistic people often understand each other relatively well thanks to our shared bottom-up processing (cf also Haker et al., 2016) – commonly represented as 'extreme attention to individual details' or a 'a cognitive/perceptual 'style' favouring detail-oriented cognition' (Valla & Belmonte, 2013), or local processing bias over global processing/holistic stimulus processing (Stevenson et al., 2018). This way of processing is commonly represented as different to non-autistic processing, where non-autistic people usually show an a priori top-down bias (Haker et al., 2016), automatically combining congruent and incongruent cues into coherent wholes, or with other words a compulsory process of forcing stimulus into a preconceived conceptual framework. In addition to individual autistic people not having their thinking processes governed by pre-existing conceptual frameworks, autistic people sharing a space generally follow each other's thinking on its own premises rather than forcing it into a conceptual framework (Seng, 2019). This is in line with Sinclair's (2010) shared autistic space.

Following Seng, we depart from the point of view that writing is part of a larger process of knowledge production and cannot be entirely separated from the cognitive processing that proceeds it. Autistic voices, if heard, may add nuances and contribute with unique perspectives due to the differences in processing. Apart from being a minority, operating in a society shaped by and for the majority, the different functioning itself may further complicate our voices being put into words, which makes the issue of autistic writing an important issue for knowledge production about autism.

Methods

The analysis in the chapter is based on two sets of data: 11 written individual narratives concerning personal experiences of writing processes, and metanarratives evolving around the process of writing this chapter itself.

Production of the individual narratives

During a workshop on writing processes at an autistic conference in Sweden, run by Hanna supported by Serena, all participants were divided into groups of three persons each. Covering three main themes (What kind of texts do you write?, When is it easy or hard to write?, How do you overcome your difficulties?), two of the participants acted as interviewer and interviewee. The third participant, 'the recorder', documented the interview. Each interview took approximately ten minutes, and after each interview, the interviewee and interviewer changed roles with each other. After all interviews had taken place, the recorder summarised the content of the interviews to the whole group of participants. This was followed by a discussion in the whole group. All 'recordings' were collected by Serena and put on a closed Facebook group for all participants to see and reflect upon.

Then different working roles were defined. Hanna and Serena were going to share the leadership. In addition to that, Hanna, Serena, and Linda were going to coach a smaller group of one to three participants in their writing processes and write metanarratives about the experience of coaching others. This division of tasks was informed by the idea that each participant in the project had different strengths and difficulties, and the aim was to support each other and combine our different strengths in the most optimal way. This included the opportunity to be active to a different extent in different parts of the writing process. All participants were given the opportunity, supported by a coach, to revise and develop their own narratives. Those who previously did the recordings were offered a chance to be interviewed within the working group. All participants were then given the opportunity to choose between contributing to the project only with data or being co-authors. Among the 11 participants, five decided only to participate with data, and six decided to be involved as authors.

Analysing narratives and production of metanarratives

During a first stage of preliminary analysis, all authors went through all data in different rounds (cf. Bertilsdotter Rosqvist et al., 2019). They wrote initial analytical reflections and suggested themes based on the data. Parallelly they wrote metanarratives around the experience of the working process at hand. They also added new insights to their own narratives of their writing processes gained from reading the others' narratives. During this process the data evolved in an open way. Two kinds of modes of writing started to become clear, not in all cases by choice. The first one consisted of singular autistic individuals writing in relative isolation, only having contact through emails and online chat. The second one was a more collective approach, where two of the authors (Serena and Dennis) met up physically and worked together. In the next step different tasks were divided between the authors; such as working on the method-section, thematising and starting to map the narratives into different themes and patterns, choosing theories and writing up the theoretical framework, including writing reflections on the theoretical framework. As in the previous stage, all authors wrote metanarratives.

At this stage it became obvious to all authors that the process resulted in a text that just kept on growing and developing. The downside was that the open process with its unclear frames/instructions resulted in anxiety and high levels of stress for all authors. We realised that the ideas of the different working roles in the process needed more time invested, before it could become efficient. Being most experienced in academic writing and sharing an interest in the development of research methods, Hanna and Linda continued working on the text after this. The authors found that they all needed to meet up physically at some points with at least one of the others to get in tune with each other and come to agreements on how to move forward. It turned out that those who lacked the opportunity to do so withdrew from the writing process.

Methodological problems

During the final writing-up process, we came to rethink the focus of the text, stressing the impact of autistic togetherness. We realised the importance of challenging the assumption of the 'lone autistic self' (writing on one's own, in a world of one's own, only with possible interaction with other people online). The autistic togetherness we propose stems from physical participation in a shared autistic space. It enables rather than disables autistic social interaction and sense of community and belonging. It stresses autistic knowledge production as being produced in social interaction and in dialogue within the physical group of autistic people. This new focus was partly empirically generated: the 'metadata' in our second data corpus involved more focus on the management of the cooperation than was originally planned. It turned out that for most participants the writing process depended on both different communication needs and the sense of community and togetherness between the writers, and that needed maintenance.

The other reason for changing focus was that during the final writing-up process Linda came to realise several methodological weaknesses with the project design. Unfortunately, these limitations remain, as the original design informed the data collection. The research design reproduced the common story about having to deal with so called deficits in isolation rather than elaborating on how to combine our strengths. Most of the data collection focused on individual experiences rather than on building something new together and studying that process as it occurred. Gathering naturalistic social interaction data from working meetings in a shared autistic space, like the workshop where it all started, and applying conversation analysis might have been a better-suited method. The management of the cooperation in the project at hand was too tentative to fully offer an alternative to dealing with difficulties in isolation or in an NT context, especially since most participants lacked opportunities to meet up for shared brainstorming, validation, and support after the initial workshop. The themes resulting from the methods at hand are valid, but they are only part of the truth.

Presentation of the participants and group of authors

All participants are from different places in Sweden and Germany. All but one have a formal diagnosis of autism. The one without a formal diagnosis is self-identified as autistic. Some also have one or more coexisting neurodivergences such as ADHD, Tourette's, or dyslexia. All have experiences of organising or being part of shared autistic spaces. All have a long experience of writing. Most write as part of their living but differ when it comes to type of writing. For all of us, writing is also part of our social life – social media writing and socialising with friends through written communication.

Findings

The individual narratives

Central in the descriptions of the writing process in the individual narratives is the double-edged situation of using 'the flow' as a major strategy. The flow, also referred to by autistic autism theorists as the 'flow state' (McDonnell & Milton, 2014) can be framed as monotropic attention on the task at hand dependent on intellectual and emotional interest-based motivation and individually adapted conditions (Murray, Lesser, & Lawson, 2005). As the text written in flow must be adapted to communicate with others and the process of writing must be coordinated in a social context (which challenges the flow), the social interaction is presented both as a barrier and as an enabler.

Following the flow: allowing ideas to take shape on their own terms

Being in and following the flow is represented as easy and fun, as something that one would rather not break from because the change in focus takes energy and building up the flow anew will take time. Being in the flow means the body and mind follow their own pace and an emotional drive without being hindered by anything internal or external. Internal hindrances may include tiredness. External hindrances may include lack of time, difficult demands, a disturbing environment, and unclear frames/instructions.

The impact of an intellectual and emotional interest in working with the text, and also the impact of time and cognitive conditions for writing, are stressed by the participants. One of them said they need an 'emotional connection to the topic in order to be able to write. It has to strike a nerve'. If that connection is not there, they get blocked and cannot write. Similarly, writing is experienced as hard when it is boring. One participant expresses a need to build up to writing, to warm up in a playful manner through different kinds of transition activities, such as 'going for a walk or doing some gardening while remaining aware that I will then write'. This can be related to Seng's claim that much of the autistic processing occurs on a level that precedes conceptualisation, or translation into words. Another participant similarly describes the forming of ideas as allowing a formless and fuzzy cloud to gradually disentangle in its own pace.

> I know that there's something intriguing going on there, but I don't know quite what. The cloud must disentangle itself first. Tentacles start to come out from the cloud. I know that they are connected to each other through the core of the idea, and I know that it will fall into place eventually, but in the beginning the connection between the tentacles is very fuzzy. Communicating the idea to others is impossible at this stage. There are too many dimensions involved and I lack the concepts.

This way of letting ideas take shape on a preconceptual level has possible connections to creativity. The continuing narrative hints at that:

> As what I'm doing is academic research and not poetry, I must connect this fuzzy idea to previous research, so I start guessing what other people might have called this intersection of dimensions that I'm interested in. There is never (so far) anyone who has done something quite like that, so I always need to relate to various research traditions and disciplines that have very little in common. I usually end up inventing my own concepts or tweaking some concepts that have been used for entirely different purposes. If I would have started out with concepts, I would never have seen what I have seen.

One writing situation that enables the flow is being in one's own time and pace. This could be unrestricted or undisturbed time, a long starting process, or time pressure. For some participants the thinking process during writing is strengthened by lengthy sessions, 'writing marathons', which requires time for recovery. For others it is important to have shorter writing sessions. The narratives about pace are also associated with having time to process and not being disturbed during the writing process, for example by being demanded to leave a text to a reader for feedback/supervision before it is finished. That is described as a result of the bottom-up processing – the main points cannot be discerned until the end. Time is also associated with executive difficulties regarding planning, assessing time for doing tasks, and getting started. This leads to stress, especially if combined with procrastination. Energy levels are hard to predict, which renders time management difficult. While for some participants time pressure is a barrier, for others it is a tool to manage perfectionism or difficulties with motivation without an external pressure.

The importance of an emotional connection to the topic is being expressed in terms of writing as a way of both processing emotions and reflecting upon a previous event. There is also a pleasure in finding the right formulations and a good structure for the text. One participant describes enjoying 'the tones and rhythms of words, creating impressions and meanings with them'. In this case, the writing can be seen as a way of supporting a bottom-up processing of details into wholes, into a more linear-chronological form. Following Seng (2019), it can also be interpreted as part of a translation process from perceptual processing to a verbal conceptualisation, where the auditory thinking style is supported by a musical approach to language whereas the visual thinking style is supported by visual metaphors, facilitating the translation.

Being in the flow is on one hand pleasurable; going into oneself as a writer, the joy and the pleasure of being swallowed up in details, developing them and following them, which can result in new discoveries and new patterns. At the same time, an awareness is expressed that this detail-focused writing process can be perceived as unnecessary and complicated or that the writer is 'bad at seeing the relationship of the details to the whole picture'. This awareness of one's own writing process can be understood as an expression of internalised cognitive ableism (cf. Carlson, 2001), part of a process of upholding the distinction between cognitive normates and others, where one's own way of writing and processing is described as lacking in relation to an imagined cognitive normate other writing process, characterised by a more NT conventional selective focus ('a holistic approach') or linear writing:

> I can get incredibly fascinated by a train of thought and want to elaborate on it, follow threads to see where they lead even though they may not be relevant for the bigger picture. A person with more of a holistic approach would probably see immediately that a track is irrelevant and be content with a footnote. On the other hand, there might be benefits in taking these detours, after all, as they may lead to unexpected discoveries that I can use later on, even if they strictly speaking fall outside the scope of the disposition.

The process is depicted as 'chaotic and non-demarcated which allows me to see what my colleagues do not see':

> I see so many nuances and so many connections. It is sort of non-linear for me. It branches out in all different directions and I see that all of these directions are important, see how they relate to each other in complex ways that cannot be forced into linearity, and yet you must write in a linear way. It is so frustrating! I see way too much. I'm drowning.

Framing the 'flow text'

The flow writing itself can be a challenge. Flow writing means that even the analytical process where detailed data are to become a coherent whole is dependent on the flow. In the second step of the writing process, the unsorted creative, inspired dump of a text, the 'flow text', needs to be managed and structured to fit it into the frame that conditions the text, such as target audience or format. The work process in the second step is described by some participants as emotional, tiring, and as requiring a certain pace. Regular recovery breaks are needed to process the details into a whole. Some participants describe this process as creating a patchwork of the text:

> Then I must create a patchwork of tiny pieces and doing that is rather tiresome. It can be difficult to deal with so many details in my brain and transforming them into a 'bigger picture'.

This can be done in two different ways. Some create the patchwork mentally and create their structure of the text within their mind. Some visualise the structure, for example through mind-mapping, or through moving around the text physically. While the flow text consists of many details and threads, it is at a later stage translated into a new whole. This is done with various strategies for structuring, limiting, erasing, sorting, and rebuilding the text. The final patchwork is more adapted to expectations of the text and text structure in a specific context, including being a linear text with a common thread and a clear end. This may pose challenges:

> For a long time, even as a researcher, I had trouble seeing the point of writing a concluding discussion and presenting conclusions at the end of a conference presentation. I thought that since the readers/listeners had followed my path all the way there, they might as well draw the conclusions themselves.

The process of going from detail to whole is described as an emotional sorting: cutting away nuances prompts a need to manage emotions associated with it:

> I prefer writing in a context of unlimited time and space. I need much space in order to maintain a common thread, because I find 'everything' important.

> Initially, it is difficult to discern what is important, as I love details. In order to avoid separation anxiety regarding elimination of text, I have a trash file where I lay aside everything that isn't put to use in the text, so that I don't need to throw it away.

The finished whole, with all details cut away, may feel false:

> When I express the world in words, I often have the feeling that I'm distorting my own autistic gaze and to some extent 'half-lie' in order to have the slightest possibility of being understood.

Here, too, writing takes place in relation to an imagined cognitive other writing process, where one's own writing process is described as something that the participant tries to 'get rid of'. The importance of creating strategies that help sift out what is most important in the text is stressed, while also highlighting the strengths of this way of writing. For example,

> trying to get rid of this way of writing because it has led to the production of much superfluous material that doesn't come to use, but sometimes when I'm writing I feel that 'something about X is missing here' and then I can go back to the trash file and maybe there is already a complete paragraph about precisely what is missing.

Self-management of cognitive processes

Since the interview themes included difficulties and strengths in the process and how difficulties are being overcome, a particular focus in the narratives concerns different cognitive challenges commonly associated with autism. Among such mentioned by the participants are managing energy levels and overload, executive impairments, and difficulties associated with co-existing divergences. All the narratives stress the importance of getting to know oneself and one's difficulties and finding out individually adapted strategies to manage them.

Some of the difficulties in need of management were related to the framing of the task; for example, when writing together with a person with a different cognitive style or when one needs to adapt to a text format that does not fit oneself, or when the frame or context of the work is unclear. When the text format is fitting, though, it supports one's writing. As another example of the impact of internalised cognitive ableism, several participants express a fear of not doing it right, expressing themselves in the wrong way, and being misunderstood. These problems may result in difficulties in initiating or maintaining the writing process. For example, one participant writes:

> Me and my supervisor [for essays] rarely have the same view of when things should be done. He may want me to attend to analysis, for example, when

I'm in the middle of writing the methods section. Then it is difficult for me to change the topic, and I may need an entire day to process having to stop writing something that I wasn't done with.

As an illustration to how the target audience may work as a supporting context for the writing process, one participant writes:

As a doctoral student I had to get used to delivering drafts and unfinished texts and receiving critical feedback along the way. Over time I learned to appreciate it – being able to receive input from others when choices had to be made and when I had a hard time finding my focus. By telling others about a project, it becomes clearer to me what I want and what I'm able to do.

The target audience may also be present indirectly, through different examples of similar texts allowing the participant to get a sense of how their text may look in the end. One participant even describes editing their texts in public spaces to get a feeling for other people's rhythms of thinking and ways of communicating. At the same time, a genuine interest in catering to the audience is also central. One participant describes how uncertainty about the reader's interest or perspective may be an obstacle in the writing, which may result in them not getting into the flow but only writing a few sentences in a day.

The metanarratives: towards an autistic togetherness

The first phase of the analysis process was marked by confusion and stress – ambiguity about the subject and the working process. Several authors note in their reflections that they needed a clearer sense of both the aim and the working process. One of them expressed that it was initially 'somewhat like moving through an immense labyrinth without seeing'. At the same time, the sense of recognising oneself in others was found rewarding. One of the authors noted that when they got to read the other participants' narratives or metanarratives, they got inspiration and motivation to write their own ones. Reading others' narratives also led to insights helping to overcome obstacles in their own writing which were made clearer during the process. One of the authors saw the work with the reflections as a tool for 'documenting exactly what is hampering me, executively, and what strategies will enable me to move forward'. This included hoping that the autistic cooperation and the setup of the project would support them to find new tools for the organisation of their own writing in the future.

The confusion and stress were due to several elements, one of them being a lack of synchronisation between the leaders. As Hanna and Serena were both busy with other projects, they were not 'simultaneously available for full engagement with this project'. We gained insights about the importance of starting and managing the process properly.

The role as coach entailed uncertainties regarding expectations. The geographic distance between the participants meant that communication had to be done from a distance. There were uncertainties as to what shape the interaction and the cooperation were to take. One of the coaches worried about the dynamics of their small group, disconnected from the whole and with asymmetric roles. In the beginning it was also not clear to what extent each participant had been informed about the setup. The flexibility of the format meant that there were many decisions to be made initially. Needs for clarity, sense of control, and knowledge about everyone's different needs are stressed as important in order to be able to function properly as a coach. For one of the authors the uncertainties brought their own difficulties to a head and made them blatantly visible. Uncovering the difficulties in their writing, closely tied to their professional identity, they felt exposed and vulnerable. This created a resistance, which in turn reinforced the difficulties. The author describes employing various strategies to get started with their assignments. However, initial uncertainties were being sorted out with straightforward communication, which created a sense of security. Findings indicate that the distribution of roles might be helpful with increased communication and more opportunities to meet, with regard to executive functioning and for social reasons: group solidarity, joint elaboration on ideas and perspectives, social accountability, and faith in the combined skills of the group.

The challenges brought out by the work process when left to their own devices made one of the authors reflect on what they had said in their narrative, about their difficulties in limiting their scope as everything is connected 'in a very non-linear way'. That contributes to their strong performance anxiety and tendency to procrastinate. In their continuing reflections, the author experimented with describing their own behaviour straightforwardly as it unfolded in the writing process. That experiment unsparingly laid bare how the performance anxiety generated a pattern of avoidance. Paradoxically enough, this generated 'ridiculously large amounts of data – bizarre what masses of text can be produced about one's own writer's block', as they comment. Even the reflection document transformed into a means for avoidance as they used it to avoid the task of structuring the main narrative.

A central theme in the author's reflections throughout the whole writing process was the need for meta-communication and candid and earnest communication about personal needs, strengths, and weaknesses, thereby enabling each other in the group to perform at their best, something that one of the other authors also emphasises in their reflections.

As a contrast to this lonely writer's difficulties to stay with the task, two of the authors, Serena and Dennis, had the opportunity to meet physically and work together from the beginning and had much less trouble. During a previous collective writing process, working together with the editing work on a book, they developed a four-step routine for writing together, with the first and last steps involving physical meetings whereas the middle steps involved individual writing and critically examining each other's texts. The first meeting involves shared brainstorming, in which both validation and straightforward criticism are central features that require a face-to-face dialogue. The meeting allows the authors to take each other's perspective and challenge their individual lines of thoughts:

We easily follow each other's thoughts. It's like a dance.

The authors describe taking delight in the process of gradually building up a clearer structure together, based on the various thoughts that have come up, before writing individually – outlining a skeleton of how the ideas fit together, which helps with the execution of the individual assignment. They also report benefitting from straightforward criticism as they edited each other's texts and met again to elaborate on the feedback, discuss the texts and make final decisions together. Building on each other's ideas, they were able to come up with something more evolved than what they would have accomplished on their own. The fact that they are each autistic was emphasised as a decisive factor because they both knew that the quality of the text was in focus rather than social games. It is worth noting that the reflections from this cooperation initially were very scarce, reporting 'no complications' in the metanarratives, whereas others' reflections on difficulties took up much more space. Prompting was required in order to get more detailed reports on what made the writing process run smoothly in these instances.

As the writing proceeded, the importance of the autistic space became clearer and clearer. One of the coaches lost their group, because their participant decided not to continue in the project as author, which was not communicated to the coach. Compared to in the start-up, the information was also scarce in general. As the other coach's working group (consisting of three other authors) was still intact, that coach (Serena) took it upon themself to send out some further information and a collection of relevant data files to all authors. They regularly checked in on the authors to see how their work was going, gave feedback, and suggested deadlines. This push was appreciated by all authors. One of the coaches noted that the personal accountability between coaches and other participants (as well as between coaches) should have been clarified for each phase and agreed upon for it to serve its purpose. All three coaches were able to communicate well about this clash and about their respective ways of functioning, and one of the coaches described how this restored their motivation. The coach also reported having learned from this that they need to meta-communicate more about their need for framing and communication. An interesting part of the process was having the space to investigate what makes one lose direction without having to worry about things getting awkward or prestige being at stake in a face-threatening way. As the coach says, while also challenging cognitive ableism,

An important take-away from this conversation is that one should not need to apologize for one's functioning. It is better to see what consequences it may entail and learn from that.

In the following interactions, the coach could take on the role of prompting instructions and deadlines from another coach a in a straightforward manner. The coach reports relief from having this seen as helpfulness rather than as being annoying:

> I am good at seeing where there are uncertainties and gaps to be filled, and it feels good to be able to contribute with that.

Yet, new problems arose as Linda and Hanna were trying to put together the final text over a geographic distance. The division of labour diverged from initial plans, and with Linda contributing more than was originally planned, they felt that they would have preferred more of a say in the design and the process of analysing from the start. The anchor to a larger collective of autistic participants had been dropped. Linda expressed her frustration in writing to Hanna. To Linda's great relief, Hanna agreed with the criticism and suggested that they meet to talk this through. They did, and during the meeting they decided to frame the text as a story about an explorative experiment opening for further questions rather than as a complete analysis of the original idea.

Concluding reflections

Each in its own way, the two data sets reflect the tension between the need for individual adaptions and a pace of one's own on the one hand and on the other hand the need for connection and shared perspectives – with the audience, but also and perhaps more importantly with an autistic collective. The individual narratives put much emphasis on individual participants' self-knowledge and conscious self-management of difficulties associated with cognitive styles, selective attention, and executive functions. This reflects the individually focused research design. The metadata also emphasise a need for a more collective management of difficulties to make use of everyone's strengths.

This tension can be interpreted as an expression of a conflict between the implicit assumptions of a sole autistic self, permeating the research design, and the notion of autistic togetherness (Sinclair, 2010) that turned out to be central. The latter stresses the need for doing things together (Bertilsdotter Rosqvist, 2019) in shared autistic spaces. This means enabling autistic sociality through awareness of the impact of context and pace for autistic communication, thinking, and writing. It also means recognising autistic social interaction as an intimate part both of forming the autistic togetherness and of writing together.

Knowledge production in research taking the sole autistic self for granted as research object is limiting and needs to be reconsidered. When creating the 'sole-autistic-self'–focused narratives, we fell into the trap of the medical perspective of autism. The results from them resemble results from traditional individual interview data in autism research and do not reflect the liberating discussions we had at the conference where the project started. One important take-away from this project is that even when conducted by autistic researchers, research that portrays autistic people in their individual bubbles can only reproduce certain results. It keeps (re)imaging the sole autistic self while failing to allow for differences in pace and create context for the meeting that would enable autistic togetherness and possibly a writing process less hampered by difficulties, including contexts for challenging cognitive ableism. This chapter is therefore partly the story of a failure, but it also generates thoughts about why the stereotypical image of autistic people as struggling loners comes up again and again. The research context creates it by focusing on the individual separated from their social context.

The findings from this enterprise still say something about the translation process that seems to occur when autistic thinking collides with the limitations of language and the frames provided by text formats and in coordination with other people in time and space in the writing process. It is important to acknowledge the existence of this barrier and the draining work that autistic people do to overcome it. Despite the limitations of the design, the findings also tentatively illustrate the need for writing processes rooted in shared autistic spaces based on autistic sociality. What we have yet to see is the development of research methods that successfully enable that kind of space throughout the research process and what would be the outcome of that.

Much could have been done differently in this project to come closer to that vision. A project infrastructure allowing continuous group discussions and access to each other's reflections rather than assigning Hanna the role of the spider in the web might have increased the sense of community in the group as a whole and enabled participants to build on each other's ideas more dynamically and make decisions together. That would have required a great investment in meta-communication at the start, to find formats suitable for the individual needs of all participants, and possibly a great deal of travelling. When designing the project, Hanna thought it would be convenient to let everyone 'do their own thing' as a way of meeting everyone's different needs. In doing so, however, the needs associated with a shared autistic space were neglected and the notion of the sole autistic self was reinforced.

The metanarratives reflecting on the ongoing process as it unfolded complemented the individual retrospective narratives by generating descriptions that were situated in a concrete context in the present. If the research design had involved more cooperation, these data would probably to a larger extent have described autistic cooperation. However, they involved methodological difficulties of their own. Different authors generated reflections in very different formats and scope, with varying degrees of detail, concretisation, and contextualisation, which rendered comparisons difficult. We also noticed that there was a tendency for authors to see problems as more noteworthy than processes running smoothly, wherefore follow-up questions had to be made to elicit accounts of what had only been depicted in brief notes. A clearer framework for the metanarratives might have mitigated these problems.

A better option would have been to record work meetings, brainstorming group discussions, and data sessions to access autistic collective writing and knowledge production as they occurred in context and in social interaction (cf. Örulv, 2008, 2012). That would possibly have enabled access to more of the translation from perceptual to conceptual thinking (Seng, 2019) as well as the autistic intimate cooperation in a shared autistic space (Sinclair, 2010). The downside is that recording the social interaction might have hampered the process and led to feelings of vulnerability and exposure.

In developing methods for research in the autism field it is important to bear in mind that autistic people are a highly discriminated group, subjected to harmful cognitive ableism. Exposing one's divergence from what is considered normal involves great risks. Recording the social interaction between autistic participants may be intrusive. It cannot be done unless sufficient trust has been built and participants have gained enough confidence with regard to their contribution. The latter may require organically developed in-group knowledge of how competences can best be combined to build on everyone's strengths. Therefore, it might be preferable to depart from already established autistic groups in shared autistic spaces if possible.

This actualises the problem that the availability of shared autistic spaces is scarce, with limited access for outsiders. Even autistic researchers may be met with suspicion due to negative experiences from autism research and public discourse in general. The presence of non-autistic researchers may transform the autistic space into NT space, as the autistic space is fragile to the neurotypical gaze. As Sinclair (2010) writes, a majority of autistic people in a space does not make it 'autistic'. One single neurotypical individual can still take 'charge of creating structure and setting the agenda' (Sinclair, 2010). We authors have all witnessed this happening and felt the discomfort as the autistic sociality collapses with the sudden subordination of the autistic participants to dominating norms that fail to take needs of autistic people into account. The vulnerability of the shared autistic space may also be due to differences in communication and processing, where the bottom-up processing allows for a well-developed brainstorming capacity among autistic people, and an ability to share and openly follow each other in the flow, where NTs' top-down bias or compulsory process of forcing stimulus into a pre-conceived conceptual framework, reusing concepts, are disturbing the flow (cf. Seng, 2019). Even in strictly autistic spaces, the ever present (albeit physically non-present) cognitive normate other and cognitive ableism casts a normative shadow over autistic ways of being in the world. In our findings this is illustrated by constant comparisons to NT ways of doing things and struggling to adhere to those norms and expectations. Creating the kind of space needed for autistic voices to be heard in research is thus as difficult as it is important. Collective autistic writing projects may enable other forms of knowledge to emerge in the field of autism research, increasing the possibilities for autistic people to become subjects rather than mere objects in research.

References

Bertilsdotter Rosqvist, H. (2019). Doing things together: Exploring meanings of different forms of sociality among autistic people in an autistic work space. *Alter; European Journal of Disability Research; Journal Europeen de Recherche Sur le Handicap, 13*(3), 168–178.

Bertilsdotter Rosqvist, H., Brownlow, C., & O'Dell, L. (2013). Mapping the social geographies of autism – online and off-line narratives of neuro-shared and separate spaces. *Disability & Society, 28*(3), 367–379.

Bertilsdotter Rosqvist, H., Kourti, M., Jackson-Perry, D., Brownlow, C., Fletcher, K., Bendelman, D., & O'Dell, L. (2019). Doing it differently: emancipatory autism studies within a neurodiverse academic space. *Disability & Society, 34*(7–8), 1082–1101.

Brodski-Guerniero, A., Naumer, M. J., Moliadze, V., Chan, J., Althen, H., Ferreira-Santos, F., Lizier, J. T., Schlitt, S., Kitzerow, J., Schütz, M., Langer, A., Kaiser, J., Freitag, C. M., & Wibral, M. (2018). Predictable information in neural signals during resting state is reduced in autism spectrum disorder. *Human Brain Mapping, 39*(8), 3227–3240.

Carlson, L. (2001). Cognitive ableism and disability studies: Feminist reflections on the history of mental retardation. *Hypatia, 16*(4), 124–146.

Haker, H., Schneebeli, M., & Stephan, K. E. (2016). Can Bayesian theories of autism spectrum disorder help improve clinical practice? *Frontiers in Psychiatry, 17*(7), 1–17.

McDonnell, A., & Milton, D. (2014). Going with the flow: Reconsidering 'repetitive behaviour' through the concept of 'flow states'. In G. Jones & E. Hurley (Eds.), *Good autism practice: Autism, happiness and wellbeing* (pp. 38–47). Birmingham, UK: BILD.

Murray, D., Lesser, M., & Lawson, W. (2005). Attention, monotropism and the diagnostic criteria for autism. *Autism, 9*(2), 139–156.

Örulv, L. (2008). *Fragile identities, patched-up worlds. Dementia and meaning-making in social interaction.* (Unpublished doctoral dissertation). Linköping University, Sweden.

Örulv, L. (2012). Reframing dementia in Swedish self-help group conversations: Constructing citizenship. *The International Journal of Self-help and Self-care, 6*(1), 9–41.

Pellicano, E., & Burr, D. (2012). 'When the world becomes 'too real': a Bayesian explanation of autistic perception. *Trends in Cognitive Sciences, 16*(10), 504–510.

Seng, H. (2019). *Annäherung an ein autistisches Erleben: Eine Collag.* (Unpublished doctoral dissertation). Martin-Luther University, Halle-Wittenberg, Germany.

Sinclair, J. (2010). Cultural commentary: Being autistic together. *Disability Studies Quarterly, 30*(1).

Stevenson, R. A., Sun, S. Z., Hazlett, N., Cant, J. S., Barense, M. D., & Ferber, S. (2018). Seeing the forest and the trees: Default local processing in individuals with high autistic traits does not come at the expense of global attention. *Journal of Autism and Developmental Disorders, 48*(4), 1382–1396.

Valla, J. M., & Belmonte, M. K. (2013). Detail-oriented cognitive style and social communicative deficits, within and beyond the autism spectrum: Independent traits that grow into developmental interdependence. *Developmental Review, 33*(4), 371–398.

Chapter 11

How individuals and institutions can learn to make room for human cognitive diversity

A personal perspective from my life in neuroscience

Matthew K. Belmonte

The peculiar challenges on the margins of the autism spectrum

My experience of science and scientists is that of an infiltrator

I am brother and uncle[1] to two people with autism, and I am a neuroscientist who studies autism. That feels awkward to say, because for so long stakeholder families and scientists have been pitted against each other: my family's experience with medical and scientific authority began half a century ago with a psychiatrist asking my mother, 'Mrs Belmonte, don't you feel *guilty*?' I grew up seeing my brother and my parents contort their lives to follow research protocols that advanced scientific careers more than scientific understanding. I've seen families' reports of disrupted sleep, gastrointestinal distress, and immune disease ignored for decades because autism was a disorder of social cognition. I've seen families' stories of heightened affective empathy and emotional sensitivity dismissed because people with autism are impaired at (cognitive) empathy. I've seen case reports of cognition and communication dismissed because people with autism, whose cognitive, perceptual, and motor dyscontrol means that they can't look and think and do at the same time, 'aren't even looking'. I've watched queues of students every September enter the medical school office through one door and emerge through the other wielding a white coat, a stethoscope, and an attitude. And I've seen families spend desperate money on quackeries from auditory training to immunoglobulin infusion because science was ignoring them. So it is hard to see myself as a bona fide member of this opposite camp, and instead I tend to think of myself as a highly effective infiltrator, an impostor who's gone all the way. This chapter, then, is my report from the inside.

Human differences in cognition, perception, and action

Part of what has made me feel so different and alien has been the traits that my brother and I share. There are many of us autism family members who as children had trouble with loud sounds, were fascinated by sensory patterns, lined up our toys in order of size or colour, had a nervous habit of hand-flapping, were clumsy, awkward, and uncoordinated, couldn't immediately recognise new faces, and felt anxious looking into others' eyes. Decades before the scientists began to recognise what came to be known as the 'broader autism phenotype', which runs in autism families (Piven, Palmer, Jacobi, Childress, & Arndt, 1997), and before autism's strong genetic basis (Bourgeron, 2015; De Rubeis & Buxbaum, 2015; Ramaswami & Geschwind, 2018) became known, I knew that my brother and I shared a way of seeing, a frame of mind that those around us just didn't get. The high-pitched hum of the flyback transformer and its changing interference pattern as I moved my head in front of the new colour television set, the twang of the springs underneath the living-room chair as I released them at different displacements, the pattern of depressions made in the earth by raindrops underneath the playground swing, such phenomena fascinated us because of their tractable, repeatable cause and effect and their parametric variability. These traits in the social context of the 1970s were described only as shyness and peculiarity; not till the 1990s did Asperger's become part of an ever broadening (Rødgaard, Jensen, Vergnes, Soulières, & Mottron, 2019) autism rubric. I argue that because autistic traits arise as a continuum throughout the population, and because the autism label is determined in part by these traits and in part by their interaction with culture (Belmonte, 2011), the question of how to accommodate autistic neurodiversity is essentially inseparable from the general question of how to accommodate human cognitive diversity. Others, too, have wrestled with the tension between the categorisation inherent in a label and the dimensional variability within and beyond its boundaries (Beardon, 2017, p. 8). This continuity between the autism spectrum and general population variability of human brains, minds, and behaviours only strengthens the imperative that every individual be approached and appreciated as an individual rather than merely as a label: if you've seen one person, you've seen one person.

Language helps us compensate and camouflage these differences

Amid all these similarities, what so fundamentally differentiated my brother and me was that I could speak. Sure, sometimes I spoke too softly or too loudly, shied away from flexible social interaction, and spent most of my time poring over science books or programming computers, but I could do these things because I was able to express language – superior language, according to my primary school teachers. My brother and I had some of the same raw material: combining it with speechlessness made it a disease, but combining it with speech made it a difference.

But successful camouflaging brings its own challenges

In citing these similarities I mean not at all to minimise or to distract from the conditions and needs of my brother, my niece, and people like them, nor – as seems to have become the vogue – to appropriate the clinical diagnosis that distinguishes them as patients and people who need treatments (Mitchell, 2019; Singer, 2019). There is a boundary between what constitutes a disease condition and what constitutes only individual difference, and I will not blur that boundary. My aim, rather, is to highlight the existence and indeed the ubiquity of individual differences in the traits in terms of which autism is defined, and in associated support needs, that might not cross the threshold for diagnosis but which do nevertheless introduce opportunities for individuals and social institutions to understand and to accommodate each other more sensitively and more effectively. These observations apply to persons who have an autism-spectrum diagnosis, persons who could have such a diagnosis, and persons who classify themselves within the 'broader autism phenotype' beyond the bounds of the diagnosis. In a very narrow way, those of us whose combinations of skills and deficits place us at or just beyond the margin of the autistic spectrum experience especial challenges: our genuine social cognitive differences are not a match for the way most social institutions work, but also are not identified and recognised and therefore do not become targets for accommodation and flexibility within such institutions (Livingston, Shah, & Happé, 2019). We tend to camouflage (Atwood, 2007; Lai et al., 2016) social cognitive differences and thus to 'pass' in social interactions, by working hard to script or to systemise (Baron-Cohen, 2008) what we're expected to say and to do. These strategies extend across the diagnostic boundary, not differing in kind for people within and outwith the autism-spectrum diagnosis (Livingston et al., 2019). And therein lies the tragic irony: time after time in the educational world, and especially the working world, our social deficits and executive and affective dyscontrol lose us the very same jobs and opportunities (Livingston et al., 2019) that our technical skills and intensity of focus have gained us. The same academic institutions that expound the scientific study of autism and individual differences are themselves hobbled by normative assumptions and prejudices, which tend to shut down and ultimately to shut out us camouflagers when we bump up against our limits (Livingston et al., 2019) at inferring organisational politics, recognising and meeting unspoken expectations, multitasking and planning for short and long terms simultaneously, and regulating our own cognition and affect in contexts of these job-related stresses. How much more we and the institutions within which we work could do for the world, if only we and they could accommodate others!

My own experiences can suggest remedies

To suggest specific avenues for such accommodation I draw inductively on my own experiences, beginning with my perspective as a school student and continuing to my experiences as a doctoral student and a junior member of faculty, during both of which recognition for scientific insights was paired with expulsion for political and interpersonal blindness. The severity of these institutional responses grew in tandem with the seniority of my position: whereas institutional and social structures afford junior scholars considerable room for naïveté, seniors are assumed to be able to manage projects and people, and in general to have mastered a host of social and organisational skills that never are taught explicitly and which in any case can be difficult to internalise. Indeed, early careers of workers with Asperger's tend to follow a strong upward arc but then to hit a wall precisely because it's at the mid-career stage that the system expects them to take on management responsibilities (Cockayne, 2015). But let us begin at the beginning, in school, because that early interaction between me and the institutions charged with my education foreshadows and illuminates all my later experiences in the world of scholarship and employment.

School

Early recognition is possible

'I think Matthew might have a problem,' the three-year-olds' preschool teacher opined. My parents were taken aback: I was the one who spoke and hugged and made eye contact and wanted to grow up to be a geologist and pestered my mother to take me to see the railway freight yard and the sewage treatment plant. My elder brother was the one with the problem; as a child I learnt to keep my head down, powerless as I was to resist my father's rage against outrageous fortune or to salve my mother's heartache. 'He doesn't play with the other children, just walks round the edge of the playground staring at the ground,' explained the teacher. So my father asked me about this behaviour. 'I'm looking for treasures,' I explained, by which I meant geological specimens. And that was that.

But camouflaged traits are overlooked

The teacher, of course, had been onto something which at the time bore no name, and thus no distinction or significance. In the twenty-first century the default conceptualisation of neuropsychiatric conditions as categorical, all-or-none diagnoses has given way to a more nuanced construal in which every individual varies along dimensions of these conditions. We all have some level of autistic traits, more or less (Constantino & Todd, 2003; Baron-Cohen, Wheelwright, Skinner, Martin, & Clubley, 2001), these traits are quite heritable, and the social communicative and repetitive behavioural aspects tend to be inherited more independently of each other than not (Ronald, Happé, & Plomin, 2005); it's when they coincide within one and the same individual that they begin to reinforce each other (Valla & Belmonte, 2013). Tellingly, my parents' reaction of denial turns out to be the norm in autism families; teachers' ratings of autistic traits in siblings of people with autism are more accurate than parents' ratings (Jobs, Bölte, & Falck-Ytter, 2018), perhaps because parents are implicitly contrasting their non-autistic children against their frankly and categorically autistic children. And what my parents didn't see, foreshadowed what the institutions around me also would not recognise.

Scientific and lexical skills can overshadow motor and social deficits, and anxiety and depression

During primary school I revealed myself as rather prototypical: impaired at motor coordination, slow in relating socially to peers, and skilful at reading, writing, and science. Although my skills eventually were recognised and catered to, nobody knew what to do with me socially, and the motor issue went unrecognised as it didn't rise to the level of clinical significance. The curriculum, and most teachers, assumed typical capacities for motor control and social cognition. I vividly recall trying to deduce the rules of kickball, which we were required to play and which everyone else on the playground seemed to know, but which never had been explained. I knew the embarrassment of being singled out, by my physical education teacher in Year 2 for being unable to coordinate arm and leg movements during star jumps, by the headmaster in Year 3 who admonished me to swing my arms when I walked instead of holding them still at my sides, by a substitute teacher in Year 4 for my chronically poor penmanship, and by my classmates in a Year 7 basketball relay race in which, unable to shoot a basket or even consistently to dribble the ball, I squandered my team's lead. This chronic mismatch with normative pedagogic assumptions about motor and social cognitive skills fed the anxiety to which I was prone and culminated in an abject crying fit in Year 7. I so loved learning, but so hated what I had to go through to do it.

Deficits can be accommodated by working with, rather than against, the phenotype

In autism itself, there are a head-on approach and an indirect approach to training social skills. Most effective can be the indirect route of developing the prerequisite skills (Karanth & Archana, 2013), including attentional, executive and motor skills, whence higher-order cognition takes wing. This approach – to reiterate – applies equally both within and beyond the categorical diagnosis of autism: too often, well-meaning schools, teachers, and families throw children into the deep end, as it were, into the midst of social-cognitively exhausting games, and into physical activities that exceed their current capacity for motor control. When I was so treated, I responded with anxiety and with consequent attempts to control that anxiety by withdrawing more exclusively into repetitive, predictable activities such as reading science fiction and programming my computer. This cognitive-behavioural dynamic again is of a piece with that in autism per se: anyone, autistic or not, when confronted with demands that tax their capacities for cognitive control and so heighten their anxiety, needs to self-regulate with repetitive behaviours that render the world tractable and predictable (Belmonte, 2008; Evans, 2000). Again, the head-on approach of obstructing that self-regulation only results in its externalisation and irruption as anger (Samson, Wells, Phillips, Hardan, & Gross, 2015) – as evidence of which my parents' house had holes kicked in its walls!

Social skills can be scaffolded by group activities of mutual interest

I found my counterpoint to the head-on approach during the summer after Year 7, in the form of a summer residential science course (Durden, 1985; Hulbert, 2018, pp. 200, 209) for which I and many similar children had qualified on the basis of standardised testing and to which we all had self-recruited. Two elements were key: at long last I had a peer group interested in the same activities as I, and we had a common activity around which we interacted socially, at a pace and with foci determined by us and in which we therefore felt invested. And the residential setting afforded many opportunities for peer interaction. The curriculum set for us was built around the concept of an 'optimal match' (Durden & Tangherlini, 1993) between each individual student's abilities and the levels of skills through which s/he was progressing; the same could be said of the social environment that we students set for ourselves (O'Reilly, 2006). This scaffolded approach is the same used in social skills therapies for autism spectrum conditions that centre on interacting with peers around an activity of common interest with well-defined rules and roles, such as Lego therapy (LeGoff, 2004; LeGoff & Sherman, 2006; Owens, Granader, Humphrey, & Baron-Cohen, 2008) or role-playing games (Rogers, 2008), and indeed most such therapeutic social activities are inventions of the participants themselves and are construed as recreation rather than explicitly as therapy. My interests and ambitions having shifted from geology through astrophysics to computer science, as I became more and more skilled at computer programming during the mid-1980s, I developed a computer-mediated peer community. Along with the interventions of many perceptive, caring, and highly able teachers who succeeded despite the school system, such foretastes of the academic and peer environments that I would discover at university sustained my hope throughout secondary school. Their passive absence from mainstream primary and secondary schooling is a missed opportunity, and even more worrisome is the head-on approach's active hostility to our self-segregation into role-playing gamer and online subcultures. Yes, we needed to develop the social skills to participate in the mainstream world, but in order to do that we needed safe spaces in which to practise those skills among ourselves without fear and anxiety.

Motor skills, likewise, can be built around shared activities

Schools' binding up of physical education with social-cognitively demanding group activities and out-of-reach motor milestones taught me, mistakenly, that physical activity was not for me. This too was a missed opportunity. It wasn't until I was in graduate school that I found that cycling was my sport. Cycling seems a common interest among many computer programmers and scientists, and I can see why: it centres on a mechanical device, involves repetitive movements, and is all about maps and wayfinding. Outside any context of team racing, it doesn't involve rapid and unpredictable demands to coordinate with other people; when riding single-file in the wind, reciprocal social interactions are manageably brief and delimited. Cycling is a perfect fit for those of us who are better at understanding regular systems than at understanding unpredictable social agents. In general, physical education curricula can involve students with such support needs by offering physical activities that, like their academic counterparts, scaffold and build peer interaction around the task rather than demanding full ability at team interaction from the very first go.

University

Proactive transitional support is needed at the beginning of university

Having slid through high school computer science, only at university was I consistently confronted with my limits and with the need to organise my studies even in this, my strongest of subjects. It was a difficult realisation, especially because my experiences of social exclusion during primary and secondary school had taught me, unhealthily, to found my self-worth on my academic ability – and once I ran up against the edge of that ability my ego became a fragile artifice indeed. The coping strategy that I had developed in school, where my social deficits always were overshadowed by my academic skills, no longer applied. When one is suddenly no longer the smartest, social deficits can fuel anxiety and depression.

Institutions should not assume that students choose to be aloof and uninvolved

Having no welcoming space to which to come home, I spent most of my waking hours in my office in the computer science building. When the halls of residence closed for the summer, I lived in the computer science building till I was discovered, then in the woods behind it. I was largely disconnected from the mainstream life of other human beings; my sustenance was books and algorithms. I moved through the campus like a neutrino, physically present but hardly ever interacting. Because it never occurred to me to consider how I might look through others' eyes, I grew a long beard and eschewed footwear; belatedly I learnt that I was seen as creepy and threatening. This lack of social perspective-taking set me apart even within spaces that ought to have been safer: in the computer science department I was scolded for abusing shared space, in the science fiction club I was irritatedly hushed for quoting lines from memory during a *Star Trek* screening. Impaired at face recognition, I either got names mixed up and committed *faux pas*, or inhibited myself for fear of doing so. Ill-timed or ill-contextualised attempts at humour either fell flat or evoked indignation. I lost interest in my engineering curriculum, and my marks were down. At a nadir, I succumbed to depression, which is common in camouflagers (Lai et al., 2016; Cage, Monaco, & Newell, 2018). It's a popular misconception that people with autism spectrum conditions aren't motivated towards social interaction; what's actually the case is that they don't display that motivation using conventional social signals (Jaswal & Akhtar, 2018), and become even more vulnerable to depression when they're excluded and ignored. This same dynamic can apply regardless of whether one has, could have, or sets oneself just outside the autism diagnosis.

Finding the right living/learning group can be the key to survival within a large institution

Ultimately it was my sheer desperation for a change that gave me the guts to try something entirely new: a residential college (Gordon, 1974, p. 236) for the creative and performing arts within the university, which had a reputation for tolerance and indeed celebration of nonconformism, was holding a banquet. I bought a ticket. If there was anywhere that I would fit in, I thought, it would be here. Nervously, I sat down at a table where I knew nobody. And they spoke with me. And they were interested. And they didn't jump to judgement. The wave of relief, the ability after so many years to exhale, to stop expending so much cognitive effort to ascertain what I *should* say and instead just to say what I thought and felt, was just like the summer science course at the end of Year 7. I had found a home. By the end of my next two years I was on the Dean's list for academic excellence, had complemented my specialisation in computer science with another in English literature, and had taken up letterpress printmaking. And without the constant work of being an actor playing myself, I was able to take up acting, writing, and directing in amateur theatre, a great activity for educating oneself about social perspective-taking (Corbett et al., 2016) and also for social scaffolding, because there's a script and one can study the characters in one's own good time, and there are light and sound boards to be operated. (This is why I've always thought that the question 'I would rather go to the theatre than to a museum' on the Autism Spectrum Quotient (Baron-Cohen et al., 2001) is an uninformative one: you'll find camouflagers in both these places, for somewhat complementary reasons.) What is it that made the difference? In a 1999 report on residential colleges (Beland et al., 1999), we described this sense of having a supportive and self-defined home as 'community-in-the-small', complementing the institutional community-in-the-large constituted by the university.

Proactive transitional support is needed at the end of university

As a student, one has a script to follow: study hard, get good marks. After graduation, the bottom drops out. My interests having grown beyond computing and beyond what I saw as all too conventional career paths, I hadn't at all explored what to do next. I landed back in London with a deep understanding of scholarly subjects but still no knowledge at all about how to apply for a job, and still unable to see myself as others saw me or to understand how others saw themselves. My strategy of cold-calling and distributing CVs must have come across as uncouth, and English employers' backhanded style of communication didn't help matters: when I was told dismissively 'Let me take your details and we shall ring you back', I actually waited all day by the phone to be rung back! Turned down for a job as a train guard on the Underground (having always been keen on trains and transport networks), I ended up underemployed as a part-time shop assistant, selling computer discs and laser printer cards, delivering computers (rather than programming them as I had done till then), and self-medicating with beer when I could afford it. As my bank balance dwindled, I bailed out and went back to the only life that I knew, returning to my old university town, squatting in a closet in my old college and working as kitchen help.

Graduate school

Without support, attachment to places and routines can impede necessary transitions

Camouflagers are, in a way, sentimentalists; because we think in terms of concrete details and places, we attach emotional significance to those places where we've been happy or content. We all too often fail to distinguish transient community from persistent geography, people from places. I have encountered many camouflagers who become stuck in a place and at a life stage; obstinately they hang around student communities as one by one their friends acquire jobs, relationships, and lives beyond university. They haunt computing societies, science fiction reading groups, theatre groups, and as their age difference grows more and more obvious, they become ostracised as creepy hangers-on, usurpers of student status, losers. The tragedy of these people is that they have so much to contribute to the world if only they could be supported into a role in which they could make that contribution. It was perhaps only by lucky circumstance that I myself avoided becoming one of these people, because doing so is the default if no action is taken. My move thousands of miles away to graduate school was another act of desperation, like my move into the residential college, because I saw what was coming if I didn't make a change and I thought that as a doctoral student I could at least continue being a student, which was a livelihood that I knew.

Residential communities are as crucial beyond as within the undergraduate years

In graduate school the intentional residential community that had been my support as an undergraduate was not to be found; I was allotted to a student flat.

Just as had happened during my first years as an undergraduate, I found myself spending all my waking hours in the laboratory or in the library, as there was little else to my life. I joined a collective of idealistic communist vegetarians (though I wasn't one) who seemed the only alternative cultural space on campus (Wolfson, 2014, p. 660). I made friends with a few graduate students who were into *Dungeons & Dragons* and *Star Trek.* The crucial difference, though, is that I never had that residential group to which to come home. Universities tend to treat postgraduates, doctoral students, and postdocs as people who bring their own lives separate from their existences as students, and indeed many do. But that one-size-fits-all assumption, that graduates will not want group living in residential, interest-focused communities, tends not to be true in the case of camouflagers, and leaves us high and dry. I have heard a researcher who studies the 'broader autism phenotype' describe participation in group houses as an indicator of non-autistic traits. On the contrary, in my own experience of fellow geeks, we are the most likely to form such intentional communities, specifically because there's no other way that we can have a community: if it isn't intentional, if it doesn't happen where we live, then it isn't as though it's going to happen in a bar. I failed where I lacked any residential community, and succeeded when I found it again in the form of a computing society (Chuu, Lei, Liang, & Tei, 2001; Gonzalez, 2019) and a group house at MIT. At Cambridge, where postdoctoral scholars were largely excluded from the residential college system, I again fell into the habit of spending most of my waking hours in the office; I remember my years there as professionally my most productive and personally my most solitary and lonely. Indeed, the twenty-first century's divorce of labour from workspace has spurred recognition of this need for community in a surge of interest in co-working (Pohler, 2012) and co-housing (Kameka, 2015; Williams, 2005), trends that can advantage camouflagers especially when focused around common interests.

Coping with job stress breaks down suddenly at one's limit of cognitive and affective control

Detail-focused, frank, and dependable, camouflagers make good employees until suddenly we don't: in the best of circumstances we can organise our own projects, remind ourselves to take others' perspectives and to anticipate their needs, and respond to changes flexibly, creating a deserved *impression of capability*, which makes us seem not to need any help or support at all. When we're placed under stress, though, our cognitively demanding approach of attending to every detail in both professional and interpersonal life overtaxes our resources (Belmonte, 2008). In this circumstance we revert to script and ritual, and as our social perspective-taking, which was never great to begin with, goes out the window we become unreasonably rigid, exacting, defensive, and argumentative. In short, we turn into obnoxious Mr Hyde. Exactly when we owe it to ourselves and to the people around us to make an effort to think and to behave mindfully, we fall back on unthinking rigidity. And colleagues don't cut us any slack, because we've already 'proven' that we don't need help. This mode of breakdown was my undoing in graduate school (Belmonte, 2017): I took up more and more of my laboratory's computer programming and systems administration tasks until, effectively, I had not only my doctoral research to conduct but also the equivalent of a full-time job. The amount of work that I was putting in didn't show, because I was good at making things run smoothly. The stress, anxiety, and insomnia came to a head around a major out-of-town conference, whence I was denied boarding on my return flight after asking an airport security guard in the United States about construal of the Fourth Amendment. The end result was my effective expulsion from graduate school and loss of priority on a publication that would have been the culmination of my four years of work. The largest failure in this tale is my own, of course, but had structures been in place to monitor my state of mental health, anticipating a breakdown of compensatory mechanisms, these bad dynamics might have been detected before all was lost. The most effective prophylaxis against this mode of cognitive-affective failure is the sort of residential learning community that I have described. Even without that secure home base, though, students and supervisors can be facilitated in structured and probing discussion about how time is being spent, what stresses are arising, and what support is needed.

In the deep end

Postdoctoral scholars need explicit teaching on how to apply for faculty posts

I was successful as a postdoc because I was so single-minded, churning out major research reports and three review papers that helped set the direction of autism research in terms of brain connectivity. But just as was the case at the end of my undergraduate years, I gave myself no time to think of where to go next. I was being invited to deliver talks at many universities and international conferences. On one such occasion my host voiced his hope that I could 'meet some people here and see some of our facilities'; it was only some years later that I realised that this was a coded invitation to apply for a faculty job at his institution, whereas I had taken it at face value and enjoyed the tour. I also discovered that many extramural colleagues had been assuming that I already held faculty rank, so many years and so many institutions having intervened since my first (failed) attempt at graduate school; as a postdoc myself, I regularly received enquiries from people who wanted to work with me as postdocs! The irony was that I was effectively unemployed, my postdoctoral funding having run out after two years, and wondering how to find out about faculty jobs. Someone told me that jobs were advertised in the *Chronicle of Higher Education*, but all that I could find there were jobs in English departments and the like. Nobody ever told me to read the *APS Observer*. And nobody ever advised me to apply for a career development grant.

The generalisable theme in this story is that camouflagers are good at following algorithms, but when nobody bothers to tell us the algorithm, the unexplained process of career progression can be as mystifying as the unexplained rules of kickball were, all those years ago. In the end I applied to the only two jobs that I found, one in an English department on cognitive literary studies advertised in the *Chronicle*, and one in a department of human development that I heard about at a conference and for which I threw together an application in a weekend, just after the deadline. I was offered both. I took up the latter, as it was at my old under-graduate institution in a town that I, again confusing geography and community, thought would be familiar.

New members of faculty need to be assessed on merits, and told what they must do to progress

The American academic tenure system is not for the faint of heart, nor the anxi-ety-prone camouflager. In the American system one gets into a faculty post by being very good at doing one thing, but one stays in a faculty post by being good at multitasking ten things at once, not to mention navigating unspoken workplace politics. In the British system one gets into a faculty post by being good enough at doing one thing to get shortlisted, and then saying the right things during a perfunctory and adversarial farce of an interview for which only the neurotypical side of the table knows the script, or indeed knows that there is a script. Autistic traits thus make people good at getting faculty jobs but lousy at keeping them in America, and good at getting shortlisted but nearly unable to get offers in Britain. It doesn't have to be this way: if candidates with individualised support needs were assessed on their research merits, and if faculties could separate research and teaching tasks into different time periods, or even into separate jobs, then getting and keeping a job after a postdoc would be straightforward. What if there were a path that would allow us to keep doing what we're good at, to keep focusing on one project at a time?

Expectations ought not to be left unspoken

As a new member of faculty, I was given a budget to set up an electroencephalog-raphy (EEG) laboratory. In my detail-orientated mindset, I gathered all possible data on all possible EEG systems. I visited manufacturers. I spent a year and a half selecting and integrating what I needed to make the best laboratory possible. What didn't come through to me from the outset is that it wasn't going to be just my laboratory; it was going to be a resource for the department and I ought to have been spending no more than a few months putting together a laboratory that was good enough and then getting on with it, making my progress visible, and cementing collaborations with the senior faculty who would be voting on my tenure. I also was drafted to be part of a team applying for a grant for a mag-netic resonance scanner; I participated but made no attempt to hide the reality that my own priorities lay with other techniques of brain imaging. And although I of course made a point of being on campus when I had to teach, the senior faculty's summary of how I spent my one non-teaching term is telling: 'Chair rightly assumed that, being a full-time faculty at Cornell, Matthew would reside in Ithaca. Matthew, however, did not interpret it that way, and believed that residence in Ithaca was not required.' I was working all the time, but this work habit of mine was not visible. Had I been mandated regular hours on campus in very explicit and concrete terms, and been given an unequivocal schedule of research responsibilities, I would have conformed. Instead I must have come across to casually observing colleagues as someone who lacked enthusiasm for the team and who wasn't working at all hard.

Multitasking must be managed so as not to exceed one's limit of cognitive control; close advice, assistance, and monitoring must be provided when circumstances preclude a focus on only one task at a time

Any new faculty member will be familiar with the impossible simultaneity of setting up a laboratory, writing up one's completed research, conducting new research, applying for grants for future research, and planning and delivering several courses for the first time. I threw myself into developing a state-of-the-art seminar on autism. The evaluations were bimodal: those who liked it gushed that they'd learnt how to read the primary research literature, had understood that scientists can argue ideologically over how to interpret their data, had had opportunity to evaluate science critically, and had received detailed and personalised qualitative evaluations. Those who hated it complained that there was no textbook, that the articles that we read disagreed, that the daily essay assignments were too much writing, that qualitative feedback prioritised detail over timeliness, and that they hadn't been given quantitative grades on formative assignments. Some students also observed that the agenda seemed preoccupied with methodological detail at the expense of thematic understanding, that feedback likewise criticised shortcomings of detail without praising students' understanding in general, and that I gave short shrift to comments for which I wasn't prepared: 'Belmonte needs to respect unexpected class comments and allow discussion to develop around them rather than spouting his ideas and condescendingly shooting down anything that surprises him'; when viewed in autism's frame of preoccupation with the detailed and the concrete, these criticisms hit home. The worst of it was that between the laboratory setup, my grant application, and my seminar course, I left myself without enough time to plan the introductory neuroscience lectures that I'd been assigned for the next term. In that disaster of a course, some days I'd walk into the lecture theatre without enough slides to fill the hour, or with someone else's slides from the previous year that I hadn't had time to review. I became a much worse teacher than I had been, and the students were short-changed. Had I been advised to use my teaching reductions from the very beginning of my employment, so as to be planning no more than one new course at a time, and had I been given intermediate deadlines for having course materials prepared for review by a mentor, I and my students might have fared better.

Students (and colleagues) need warning not to misinterpret vocal and motor traits as aggression or as condescension

Some of my students' evaluations called me angry, rude, and unapproachable, and it was only belatedly that I recognised that, just as had happened with that airport security guard, my loud, affectively dysmodulated voice and stance when I was excited about making a point had been mistaken for an angry voice and confrontational body-language. Several students also called me condescending, jolting me into a realisation that when I focus on delivering and discussing the content, I cannot simultaneously focus on adjusting my vocal tone and pace to the speaker, and therefore risk being perceived as lecturing and pedantic. When I next taught a seminar, I included on the syllabus an explanation of these behaviours of mine and how they relate to autistic traits. (This context was all the more pertinent because the subject of the seminar was autism.) The allegations of anger disappeared, though some students still complained of perceived condescension.

People who lack social insight are liable to credit poor judges; they must be confronted with those who would be their most severe critics in the workplace, but also supported with knowledgeable impartial advice

The academic department that had employed me was, not unlike many others, riven with political undercurrents and cliques of different scholarly emphases and agendas. I stepped into the middle of this unspoken factionalism under the false assumption that because it was the department as a whole that had hired me, each member of the department would harbour the same expectations of me. In retrospect, those with other agendas withdrew from advising me whereas those with related agendas harboured a Pollyanna faith that a positive tenure decision for me was assured. Had I had a group of intramural mentors spanning all the factions, my eyes might have been opened. My main source of external advice at the time was a friend from the *Dungeons & Dragons* group who had dropped out of graduate school and, unbeknownst to me, fallen into alcoholism; she too had been cheated by the academic system's lack of advising and by the unspoken and ill-defined nature of its expectations, and she had left without finishing. From time to time she would telephone me and try to convince me that my life would be ruined unless I quit my job. Perhaps because she had been a senior graduate student at the time that I was beginning graduate school, and because my sense of being an impostor within academia had never disappeared, I began to listen to her, feeding an anxiety as to whether I was making the right life choice.

Once a career decision has been made, stick with it for a fixed term! Don't anxiously second-guess

With her advice weighing on me, when I was confronted with an unsolicited offer of a non-faculty job I dithered. After seven months my faculty post seemed a train-wreck of poor multitasking, denied grant applications, stagnant research, and failed teaching. If I left my tenure-track faculty job I would never get another, and I would have to work on someone else's research agenda not directly related to my own interest, autism; but I would be in a city whose geography/community I knew, and facing a stress level that would be manageable. If I stayed, I felt like I was going to explode. At the end of a dark day, with clouds closing in, a few seconds before the midnight deadline for my reply, I emailed my acceptance of the job, and began arranging a leave of absence from my faculty post. The next morning dawned bright and sunny and spring-like and I found myself already regretting what I had done, yet feeling bound to follow through. In the event, I couldn't devote full attention to the new job because I couldn't stop thinking about autism, spending my spare time on grant applications and manuscripts. In retrospect, I ought to have treated my faculty post as a commitment to a fixed term, no more, no less; that is, I ought to have stuck with it, but without any expectation of what would come next, and therefore without any anxiety about the looming tenure decision. Such commitments need to be made in advance because camouflagers' executive function makes us liable to get stuck on under-determined choices, unable to put them aside once decided. Overwhelmed by a world of what-ifs, of hypotheticals and counterfactuals, we become unable to live the real life in front of us.

Single-mindedness combined with literal interpretation can be perceived as aggression; all parties can benefit from training in mediation and conflict resolution

Even amid so many missteps and miscommunications, I might yet have attained reappointment halfway to tenure had I not taken leave to try out the other job, and were it not for a chance occurrence at that other workplace: One night I was observed via CCTV and mistaken for a thief, perhaps because I was leaving at 2 a.m. and casually attired, neither of which is unusual for me; I have a long his-tory of being approached and sometimes detained by authorities because my behaviour differs from that of most other people. I asked an administrator who it was who had been monitoring the camera because I wanted to discuss the incident with that person. She replied, 'I do not have a specific name to provide you with', though I deduced from circumstances that she must indeed have had the monitor's name. I kept emailing her; she kept reiterating, 'I have no further information for you in this matter', and eventually stopped responding to my repeated queries. When she left for a different institution in the same city, I called at her new office to try to get a definitive answer, but instead got a *persona non grata* notice and a police visit to my own department's chairperson. A friend later suggested that most people aren't calibrated socially for my level of obsessionality, and that 'I have no further information for you in this matter' might actually have been code for 'I don't want to tell you'. Perhaps, then, the administrator had thought that she had actually answered my question, and was wondering why I kept asking it at the same time that I was wondering why she wouldn't answer. Institutions and individuals can head off such misunderstandings by being open to face-to-face meetings when miscommunications seem to be persisting via email, and asking themselves to reframe each party's perspective from the other's point of view. I have found formal mediation training (MIT, 1996) of great value in framing rules and procedures for conflict resolution.

If you're a stranger, go to a strange land

Approaches to the foreign[2] are a useful model for approach to people with autistic traits

Some years ago, Gillberg, Steffenburg, Börjesson, and Andersson (1987) found that within the Swedish population, children born to immigrant mothers had a greater chance of being autistic. Speculation arose about some mysterious protective effect of Scandinavian population genetics, but the most parsimonious explanation is that autistic nonconformism with mainstream culture presents no barrier in the eyes of partners who come from outside that culture, and therefore children of immigrant mothers are more likely also to be children of autism-prone fathers (Gillberg & Gillberg, 1996). The same seems true of migrants to any social context, including the workplace: after the stress of my tenure failure it was a relief to accept a post in India where, having spent my life till that point feeling like an alien, I entered a sociocultural context within which I actually *was* an alien, where my behavioural oddities were ascribed to incomplete acculturation, and where each context's demands were explained to me. What if institutions could treat any newcomer the same as they treat foreigners; what if people whose cognition and behaviour render them foreigners within their 'own' cultures were afforded the chance to learn these cultures as explicitly as we learn foreign ones?

It was a bitter irony to be awarded a large and prestigious career development grant just a few days before the senior faculty voted to deny me reappointment halfway to tenure, a loss not only for autism research but also for me and the institution equally. The lessons for all concerned are these: although the world is, by definition, a 'neurotypical' one and, as such, individuals must not shirk the responsibility to adapt reasonably, institutions can facilitate this process by learning to accommodate reasonably. Work with, rather than against, each individual's cognitive strengths. Scaffold social skills with activities that appeal to these strengths. Support transitions, to overcome unproductive and limiting attachment to places and routines. Encourage themed living/learning groups, at all years and levels. Be prepared for sudden breakdowns of coping mechanisms at one's limit of stress and anxiety. Give explicit instructions about how to get, to do, and to keep a job. Advise, assist, and monitor to manage multitasking. Educate students, peers, and authorities about vocal, motor, and executive traits liable to misinterpretation as aggression or condescension. Integrate knowledgeable, impartial, and unbiased advising, and avoid occasions for second-guessing decisions based on such advice. And offer every newcomer the same tolerance and patient explanation that would be offered to a foreigner, because many of us are, in a sense, visitors from another world, here and moving among you.

Note

1 Editor's note: The author of this chapter has confirmed to us that his unnamed brother and niece have consented to being referred to.
2 Editor's note: The author uses the expression 'the foreign' as a general reference to foreign cultures and individuals.

References

Atwood, T. (2007). *The complete guide to Asperger's syndrome*. London: Jessica Kingsley.
Baron-Cohen, S. 2008. Autism, hypersystemizing, and truth. *Quarterly Journal of Experimental Psychology*, *61*(1), 64–75.
Baron-Cohen, S., Wheelwright, S., Skinner, R., Martin, J., & Clubley, E. (2001). The autism-spectrum quotient (AQ): Evidence from Asperger syndrome/high-functioning autism, males and females, scientists and mathematicians. *Journal of Autism and Developmental Disorders*, *31*(1), 5–17.
Beardon, L. (2017). *Autism and Asperger syndrome in adults*. London: Sheldon Press.
Beland, C., Belmonte, M. K., Cohen, A., Gratt, J., Lai, Y., Man, A., & McDougal, S. (1999). *A creative tension: The report of the Dorm Design Team to the Residence System Steering Committee on the Cambridge college system and its American analogues*. Cambridge, MA: Massachusetts Institute of Technology. URL: (Retrieved December 2019) https://web.archive.org/web/20130521080312/http://web.mit.edu/residence/systemdesign/cambridge1.html
Belmonte, M. K. (2008). Human, but more so: What the autistic brain tells us about the process of narrative. In M. Osteen (Ed.), *Autism and representation* (pp. 166–179). New York: Routledge.
Belmonte, M. K. (2011). The autism spectrum as a source of cognitive and cultural diversity. *Ranchi Institute of Neuro-Psychiatry and Allied Sciences Journal*, *3*, S46–S54.
Belmonte, M. K. (2017). My lost finding: 'This was my firstborn, and its loss still stings'. *The Psychologist*, 13 November 2017. URL: (Retrieved December 2019) https://ThePsychologist.bps.org.uk/my-lost-finding
Bourgeron, T. (2015). From the genetic architecture to synaptic plasticity in autism spectrum disorder. *Nature Reviews Neuroscience*, *16*(9), 551–563.
Cage, E., Di Monaco, J., & Newell, V. (2018). Experiences of autism acceptance and mental health in autistic adults. *Journal of Autism and Developmental Disorders*, *48*(2), 473–484.

Chuu, C., Lei, M., Liang, C., & Tei, A. (2001). *The Student Information Processing Board: The social and technical impact of an MIT student group*. Cambridge, MA: Massachusetts Institute of Technology. URL: (Retrieved December 2019) https://web.mit.edu/6.933/www/Fall2001/SIPB.pdf

Cockayne, A. (2015). How talented people with Asperger's are locked out of the career system. The Conversation. URL: (Retrieved December 2019) http://TheConversation.com/how-talented-people-with-aspergers-are-locked-out-of-the-career-system-41870

Constantino, J. N., & Todd, R. D. (2003). Autistic traits in the general population: A twin study. *Archives of General Psychiatry, 60*(5), 524–530.

Corbett, B. A., Key, A. P., Qualls, L., Fecteau, S., Newsom, C., Coke, C., & Yoder, P. (2016). Improvement in social competence using a randomized trial of a theatre intervention for children with Autism Spectrum Disorder. *Journal of Autism and Developmental Disorders, 46*(2), 658–672.

De Rubeis, S., & Buxbaum, J. D. (2015). Genetics and genomics of autism spectrum disorder: embracing complexity. *Human Molecular Genetics, 24*(R1), R24–R31.

Durden, W. G. (1985). Early instruction by the college: Johns Hopkins's Center for Talented Youth. *New Directions for Teaching and Learning*, 1985(24), 37–46.

Durden, W. G., & Tangherlini, A. E. (1993). *Smart kids: How academic talents are developed and nurtured in America*. Boston, MA: Hogrefe & Huber.

Evans, D. W. (2000). Rituals, compulsions, and other syncretic tools: Insights from Werner's comparative psychology. *Journal of Adult Development, 7*(1), 49–61.

Gillberg, I. C., & Gillberg, C. (1996). Autism in immigrants: A population-based study from Swedish rural and urban areas. *Journal of Intellectual Disability Research, 40*(1), 24–31.

Gillberg, C., Steffenburg, S., Börjesson, B., & Andersson, L. (1987). Infantile autism in children of immigrant parents: A population-based study from Göteborg, Sweden. *British Journal of Psychiatry, 150*(6), 856–858.

Gonzalez, S. (2019). *Fuzzball turns 50: Compute culture then and now*. URL: (Retrieved December 2019) www.youtube.com/watch?v=E8Fcr93c-v4

Gordon, S. S. (1974). Living and learning in college. *Journal of General Education, 25*(4), 235–245.

Hulbert, A. (2018). *Off the charts: The hidden lives and lessons of American child prodigies*. New York: Knopf.

Jaswal, V. K., & Akhtar, N. (2018). Being vs. appearing socially uninterested: Challenging assumptions about social motivation in autism. *Behavioral and Brain Sciences, 19*, 1–84.

Jobs, E. S., Bölte, S., & Falck-Ytter, T. (2018). Spotting signs of autism in 3-year-olds: Comparing information from parents and preschool staff. *Journal of Autism and Developmental Disorders, 49*(3), 1232–1241.

Kameka, D. (2015). Neurodiverse cohousing: What is it and why does it matter? Paper presented at *2015 National Cohousing Conference*, Durham, NC. URL: (Retrieved December 2019) http://oldsite.cohousing.org/node/3084

Karanth, P., & Archana S. (2013). Exploring prerequisite learning skills in young children and their implications for identification of and intervention for autistic behavior. In B. R. Kar (Ed.), *Cognition and brain development: Converging evidence from various methodologies* (pp. 145–154). Washington, DC: American Psychological Association.

Lai, M., Lombardo, M. V., Ruigrok, A. N. V., Chakrabarti, B., Auyeung, B., Szatmari, P., Happé, F., Baron-Cohen, S., & MRC AIMS Consortium. (2016). Quantifying and exploring camouflaging in men and women with autism. *Autism, 21*(6), 690–702.

LeGoff, D. B. (2004). Use of LEGO as a therapeutic medium for improving social competence. *Journal of Autism and Developmental Disorders, 34*(5), 557–571.

LeGoff, D. B., & Sherman, M. (2006). Long-term outcome of social skills intervention based on interactive LEGO play. *Autism, 10*(4), 317–329.

Livingston, L. A., Shah, P., & Happé, F. (2019). Compensatory strategies below the behavioural surface in autism: A qualitative study. *Lancet Psychiatry, 6*, 766–777.

MIT (1996). Mediation training offered. MIT Tech Talk, 11 December 1996. URL: (Retrieved December 2019) http://news.mit.edu/1996/mediation-1211

Mitchell, J. (2019). The dangers of 'neurodiversity': Why do people want to stop a cure for autism being found? *The Spectator* 19 January 2019. URL: (Retrieved December 2019) www.spectator.co.uk/2019/01/the-dangers-of-neurodiversity-why-do-people-want-to-stop-a-cure-for-autism-being-found/

O'Reilly, C. (2006). Maximising potential – both academic and social-emotional. In C. M. M. Smith (Ed.), *Including the gifted and talented: Making inclusion work for more gifted and able learners* (pp. 85–100). Abingdon: Routledge.

Owens, G., Granader, Y., Humphrey, A., & Baron-Cohen, S. (2008). LEGO therapy and the social use of language programme: An evaluation of two social skills interventions for children with high functioning autism and Asperger Syndrome. *Journal of Autism and Developmental Disorders, 38*(10), 1944–1957.

Piven J., Palmer, P., Jacobi, D., Childress, D., & Arndt, S. (1997). Broader autism phenotype: Evidence from a family history study of multiple-incidence autism families. *American Journal of Psychiatry, 154*(2), 185–190.

Pohler, N. (2012). Neue arbeitsräume für neue arbeitsformen: Coworking spaces. *Österreichische Zeitschrift für Soziologie, 37*(1), 65–78.

Ramaswami, G., & Geschwind, D.H. (2018). Genetics of autism spectrum disorder. *Handbook of Clinical Neurology, 147*, 321–329.

Rødgaard, E. M., Jensen, K., Vergnes, J. N., Soulières, I., & Mottron, L. (2019). Temporal changes in effect sizes of studies comparing individuals with and without autism: A meta-analysis. *JAMA Psychiatry, 76*(11), 1124–1132.

Rogers, A. (2008). Geek love. *New York Times*, 9 March 2008. URL: (Retrieved December 2019) www.nytimes.com/2008/03/09/opinion/09rogers.html

Ronald, A., Happé, F., & Plomin, R. (2005). The genetic relationship between individual differences in social and nonsocial behaviours characteristic of autism. *Developmental Science, 8*(5), 444–458.

Samson, A. C., Wells, W. M., Phillips, J. M., Hardan, A. Y., & Gross, J. J. (2015). Emotion regulation in autism spectrum disorder: Evidence from parent interviews and children's daily diaries. *Journal of Child Psychology and Psychiatry, 56*(8), 903–913.

Singer, A. (2019). Speech delivered at 'Clinical Strategies for Including Severely Affected Individuals in Neuroscience Studies', *International Society for Autism Research*, Montréal, 3 May 2019. URL: (Retrieved December 2019) www.ncsautism.org/blog//including-severe-autism-in-neuroscience-research

Valla, J. M., & Belmonte, M. K. (2013). Detail-oriented cognitive style and social communicative deficits, within and beyond the autism spectrum: Independent traits that grow into developmental interdependence. *Developmental Review, 33*(4), 371–398.

Williams, J. (2005). Designing neighbourhoods for social interaction: The case of cohousing. *Journal of Urban Design, 10*(2), 195–227.

Wolfson, T. (2014). Activist laboratories of the 1990's. *Cultural Studies, 28*(4), 657–675.

Challenging brain-bound cognition

Chapter 12

Understanding autistic individuals
Cognitive diversity not theoretical deficit

Inês Hipólito, Daniel D. Hutto and Nick Chown

'I will teach you differences!'

King Lear, Act I, Scene 4

Introduction

What autistic people tend to think about, the way they think about things, and the ways they interact with others is atypical when compared to the population at large. The cognitive diversity of autistic people, and its many variations, is well-documented and much discussed.

This paper has the potential to contribute to the neurodiversity movement by providing philosophically motivated reasons for thinking differently about the cognitive styles of autistic individuals. In particular, it challenges the prevalent mindreading characterisation of everyday social cognition that promotes the view that autism is an underlying condition that is best explained in terms of deficiencies in inferential capacities to form and test hypotheses.

In a recent opinion piece on the neurodiversity movement published in *Scientific American* on 30 April 2019, Baron-Cohen (2019) reminds of what he describes as the 'huge heterogeneity' among those people who fall within the autistic spectrum.

> Some autistic people have no functional language and severe developmental delay (both of which I would view as disorders), others have milder learning difficulties, while yet others have average or excellent language skills and average or even high IQ. What all individuals on the autism spectrum share in common are social communication difficulties (both are disabilities), difficulties adjusting to unexpected change (another disability), a love of repetition or 'need for sameness,' unusually narrow interests, and sensory hyper- and hypo-sensitivities (all examples of difference). Autism can also be associated with cognitive strengths and even talents, notably in attention to and memory for detail, and a strong drive to detect patterns (all of these are differences). How these are manifested is likely to be strongly influenced by language and IQ.

Undeniably, there is enormous variability in the full spectrum of cognitive styles exhibited by autistic individuals. Every autistic individual has their own distinct cognitive style, just as every non-autistic individual does.

Beyond merely acknowledging and carefully cataloguing the heterogenous cognitive diversity of autistic individuals, those attracted to the medical model have made persistent attempts to classify these cognitive styles, treating them as a part of a constellation of symptoms that are expressive of an underlying condition – a condition that is typically denoted by the labels 'autism' or 'autistic spectrum disorder'.

Over the decades there have been many attempts to understand the true character and ultimate basis of the totality of symptoms generated by the supposed underlying autistic condition from which autistic people allegedly suffer. Yet the current received view is that the hunt for a single condition that explains and accounts for the full set of autistic symptoms in a unified manner is a snark hunt. It is now widely accepted that 'no single aetiology can account for all differences in presentation' (Ure, Rose, Bernie, & Williams, 2018, p. 1068). Naturally, the idea that autism is comprised of a cluster of underlying conditions – not a single condition – lends itself naturally to classifying the heterogeneity of autistic individuals in terms of various species and subtypes of autism.

The aim of empirical research into these assumed underlying conditions that make up autism is directed at identifying specific biological markers for distinct aspects of autistic disorder – aspects which, by the lights of those who buy into the medical model wholesale, are understood to be neurodevelopmental in nature.

In line with these developments, some researchers have set their explanatory sights more modestly. They zoom in to focus only on what underpins the atypical patterns of social cognition exhibited by autistic individuals – namely, their distinct style interacting with others and their limitations in fluidly understanding what motivates actions. For them, understanding what explains the social cognitive aspects of autism alone would still constitute a major advance. This would surely be so if, as Baron-Cohen (2019) maintains, social communicative difficulties are found across the autistic spectrum and 'aspects of social cognition reflect areas of disability in autism, and are often the reason for why they seek and receive a diagnosis'.

The Diagnostic and Statistical Manual of Mental Disorders (DSM-5) (APA, 2013) concurs, taking the atypical styles of social cognition of autistic individuals to be among the most diagnostically important criteria and defining features of the condition. The DSM-5 tells us that Autism Spectrum Disorder is characterised by persistent

> Deficits in social-emotional reciprocity, ranging, for example, from abnormal social approach and failure of normal back-and-forth conversation … to failure to initiate or respond to social interactions; Deficits in nonverbal

communicative behaviors used for social interaction; Deficits in developing, maintaining, and understand relationships.

(American Psychiatric Association, 2013, p. 50)

The goal of accounting for the distinctive patterns of social cognition of autistic individuals has launched a thousand explanatory ships, all of which set forth to discover what underpins at least the social cognitive aspects of the autistic mind. This paper raises doubts about a specific class of explanations of the social cognitive styles of autistic individuals – Theory Theory, or TT, explanations.[1] TT comes in many forms – but what is common to all versions is the proposal to explain the basis of our social cognition by appeal to machinery of mind that makes use of theories of some sort that are understood to involve contentful inferential processes. TT is a worthwhile target since it is the dominant and most popular framework for answering the Explanatory Challenge of what, supposedly, underpins everyday social cognition and what goes systematically wrong with social cognition for autistic people.

Social cognition can be neutrally understood as denoting 'our ability to understand and interact with others' (Spaulding, 2010, p. 120). As such, it is important to remind ourselves that everyday social cognition can be depicted in various ways. Appropriately characterising the nature of social cognition poses a special kind of philosophical challenge. Let us call this the characterisation challenge.

Ultimately, we will argue that the various explanatory proposals of TT only look promising so long as a certain kind of answer is given to the characterisation challenge. That is to say, TT proposals look promising if we accept the dominant characterisation of social cognition – the standard mindreading story – which holds that 'in order to understand and successfully interact with other agents, neurotypical adult humans attribute mental states to other agents in order to explain and predict their behavior' (Spaulding, 2018, p. 7).

Depicting social cognition in such spectatorial terms licenses the received and longstanding view in the field that, 'many people with autism spectrum conditions have a specific impairment in mindreading' (Heyes, 2018, p. 149). As we shall reveal, it is no accident that the credibility of TT explanations of the alleged social cognitive deficits of autistic individuals depend on thinking of social cognitive styles of autistic individuals in terms of impoverished mindreading – as impaired attempts to get at the contents of other minds, that occur whenever autistic people attempt to engage in everyday social cognition.

This paper argues that the fate of *any* proposed TT answer to the explanatory challenge stands or falls with the appropriateness of giving a mindreading answer to the characterisation challenge. The first two sections provide details of the current state of the art with respect to TT proposals about how to understand and explain autistic social cognition. Section 2 focuses on old school modularist TT proposals, noting their theoretical and explanatory limitations. Section 3 examines new school predictive processing TT proposals, highlighting what has made them appear more theoretically and explanatory promising to many researchers.

Section 4 then provides a diagnosis of why we should reject any kind of TT proposal about the supposed social cognitive deficits of autistic individuals. Our concerns are about any TT proposal that takes seriously the core assumption that the primary and pervasive way that we engage and connect with others is by means of theorising. We raise objections to TT proposals as a class by providing reasons for thinking that a mindreading answer to the characterisation challenge obscures the true nature of everyday social cognition. We conclude that the crucial step of characterising the social cognitive styles and tendencies of autistic individuals as some kind of mindreading deficit is a mistake. The paper closes by encouraging the adoption of alternative, non-mindreading ways of understanding the social cognitive styles of autistic individuals. We contend that an enactivist alternative can offer an antidote to thinking of the diverse social cognitive styles of autistic individuals in terms of underlying sub-personal cognitive deficits rather than in terms of the cognitive differences of whole persons.

Old School, Mental Module TT

A longstanding, high profile TT hypothesis about what best explains the distinctive social cognitive patterns of autistic individuals holds that these stem from those individuals having a faulty or poorly functioning Theory of Mind or, ToM, module.

In general Theory Theory, or TT, proposals about social cognition assume that when we understand minds in daily life, we use the same sorts of tools that we use to understand other non-mental phenomena. That is to say, we use the same sort of tools we use everywhere in the sciences – namely, theories that aim to tell us about the unobservable, hidden causal structure of the world.

A ToM is a very particular kind of theory; it is assumed to have a distinctive sort of content. A ToM is made up of mental-state concepts that feature in theoretical postulates that comprise the core general principles of a theory of everyday psychology. The content of the ToM that normally developing humans use, so the story goes, is what enables most of us to navigate our everyday social world fluidly and with ease. We succeed in understanding others if we manage to infer their mental states correctly by applying a ToM, thereby bringing the laws of everyday psychology to bear on particular cases.

As such, ToM variants of TT hold that, for most of us, the heavy lifting in everyday social cognition is done by our acquaintance with and use of laws of governing everyday social cognition. Modularists take this idea a step further. They hold that ToM laws are housed in a special cognitive mechanism – a ToM Module, or ToMM. A ToMM is an architecturally distinct mental module – one that is solely dedicated to the special work of enabling us to predict and explain the actions of others by accurately attributing mental states contents to them. Believers in classic ToMMs assume that 'the mind contains a single mental faculty charged with attributing mental states (whether to oneself or to others)' (Carruthers, 2011, p. 1).

Those who posit ToMMs assume that however such modules are acquired, they are the means by which everyday social cognition is normally conducted by our species. Nativist accounts of ToM assume that it is a biological device that comes built-in as standard for all normally developing members of our species (Fodor, 1983). Others hold that ToMMs are acquired during ontogeny (Karmiloff-Smith, Klima, Bellugi, Grant, & Baron-Cohen, 1995). Some, such as Scholl and Leslie (1999) even propose that 'normal children seem to develop the same ToM at roughly the same time' (p. 138).

In their heyday, modularist theories of mind aided and abetted the idea that impaired mindreading abilities, rooted in damaged or atypically functioning neurocognitive machinery of a ToMM, were responsible for the social cognitive patterns displayed by autistic individuals.

We see these ideas brought together in the work of Baron-Cohen. According to its original formulation, Baron-Cohen's (1995) mindblindness hypothesis proposed that 'children and adults with the biological condition of autism suffer, to varying degrees, from mindblindness' (1995, p. 5; Brewer, Young, & Barnett, 2017; Gilleen, Xie, & Strelchuk, 2017; Livingston and Colvert, 2019). By Baron-Cohen's (2000) lights, difficulty in social cognition, understood as a mindreading impairment, is the 'core and possibly universal abnormality of autistic individuals' (Baron-Cohen, 2000, p. 3).

Summarising work in this vein, Brüne (2005) reports a range of findings that suggest to many that we find the fingerprint of an 'impaired ToM in a variety of neuropsychiatric disorders' (Brüne, 2005, p. 21). Concomitantly, modularists hypothesised that these various patterns of autistic mindblindness are caused and explained by problems with the ToMMs of people in these populations. Thus, a standard proposal in the field, even today, is that 'the functional or structural disruption of the neural mechanisms underlying ToM may give rise to various types of psychopathology' (Brüne, 2005, p. 21).

The faulty ToMM proposal contends that autistic individuals are prevented from attributing contents to other minds accurately or, in the most extreme cases, doing so at all. What makes the faulty ToMM hypothesis about the social cognitive tendencies of autistic individuals attractive to many is that

> to see a person with autism, we are told, is to see what happens to a human being when the ability to mentalize … is switched off.… On the surface, this is neatly specific … The 'theory of mind' explanation seems to fit the facts.
>
> (Belmonte, 2009, p. 121)

As traditionally understood, a defining feature of modules is that they are informational encapsulated, both vertically and horizontally (Matthews, 2019; Quilty-Dunn, 2019; Raftopoulos, 2019). Information contained in each module is vertically encapsulated from other modules and horizontally encapsulated from the information available in the cognitive system.

Modules have limited cognitive interests and concerns. In getting their epistemic work done, they operate on a strictly need-to-know basis, and they – apparently – don't need to know much. It is assumed that modules work better and faster by restricting their concerns to specialised dealings with only certain topics. It is because of their informationally encapsulated domain-specificity that they are not informed and updated by all the contentful knowledge that might possibly be communicated to them. The limited communicative repertoire of modules is the peculiar characteristic that secures their status as mental modules.

Accordingly, each type of mental module is assumed to be restricted in the subject matter of its concern. Modules are domain-specific in a sense that only a circumscribed class of inputs will activate them. It is this feature of modules that makes them dissociable such that they can be selectively impaired, damaged, or disabled without effecting the operation of other systems.

Putting all of this together, ToMM theory holds that the main work of predicting and explaining the behaviour of others by assigning mental states is done in isolation from the operation of other cognitive systems. The essential character of ToMM is that it provides specialised theoretical knowledge of its particular domain and it can function by and large independently of other modules and central cognitive processes.

Since ToMMs are dissociable components, both from other modular devices and from central cognition, impairment of a ToMM will not directly impair the functioning of other cognitive mechanisms. In conclusion, the awkward and failed social interactions of autistic individuals are thus put down to the alleged fact that they lack a ToMM or are unable to wield their ToMs well in practice so as to accurately represent the mental states of others.

It is not enough to have a working ToMM, to mindread successfully. Successful mindreading also requires being able to apply one's ToM sensitively in ways that address the particularities of specific cases. That requires adjusting for relevant differences between cases by making allowance for a range of variables including a great deal of background belief and knowledge about what the other person knows and thinks, how they are likely to react, what is the most likely way someone would react in such cases, and so on and on. In short, believers in ToMMs are obliged to explain how the core ToM we allegedly use is applied sensitively in situ (for a more detailed discussion of this point see Hutto, 2008, Chap. 1).

It is wholly unclear how having a general ToM machinery that works in isolation from relevant background knowledge could possibly enable us to cope with ad hoc details and idiosyncratic attitudes we need to cope with in each new situation. Without supplement, ToMMs would be at a loss precisely when it comes to explaining how we deal with details; it is uncomfortably quiet on how we fluently come to understand particular people and in particular circumstances. Yet, as Maibom (2009) observes 'folk psychological knowledge is knowledge of the (empirical) world only if it is combined with knowledge of how to apply it' (p. 361).

An adequate explanation of how ToMMs could underpin everyday social cognition is required if a ToMM story is to be believed about how social agents grasp these social idiosyncratic details so as to understand and come to make sense of one another.

How could a ToMM operate successfully in isolation from the background knowledge and belief that appears to be needed to inform and direct its use in specific cases? Pivotally, the supporting knowledge needed for applying a ToM sensitively cannot be supplied by a ToM and, indeed, given the contextual nature of the supporting knowledge, it isn't possible to specify it in advance at all. This reminds us that the business of socially engaging with and coming to understand others is deeply context sensitive. There are simply no algorithms with the right properties that would allow us to anticipate the relevant possibilities.

New school, Bayesian Brain TT

It is neither clear how, nor if, defenders of the classic ToMM can respond adequately to the serious theoretical and explanatory concerns raised in the previous section. In this light, a tempting way to go might be to seek to change theoretical horses by jumping on the Bayesian brain bandwagon.

The Bayesian Brain Hypothesis, or BBH, contends that cognition is, through and through, concerned with making and improving on its predictions about the causal structure of the world. Tirelessly and proactively, our brains are forever trying to look ahead in order to ensure that we have an adequate practical grip on the world in the here and now. On this view, our brains do not sit back and receive information from the world, form truth evaluable representations of it, and only then work out and implement action plans. Instead, the BBH holds that the basic work of brains is to get ahead of the curve by making the best possible predictions, in advance, about what the world is likely to throw at us. This is all part of the bigger job of cognition which, in all its varieties, is to try to get a sense of what is going on behind the sensory scenes by advancing, testing, and refining inferences to the best explanation on multiple spatial and temporal scales. Through this continual and dynamic process, so the BBH claims, we get a better and better fix on the true causal structure of the world.

Although the BBH made its name for its attempts to better account for the nature of perception–action cycles there has been a move to apply it to understand a much wider range of cognitive phenomena, including social cognition – especially when the latter is construed in mindreading terms (see, e.g. Pezzulo, 2017).

The BBH is a full-blooded type of TT. Like its ToMM cousin it seeks to give an account of the mindreading processes that it assumes lie at the heart of everyday social cognition. As noted above, the idea is that – at its core – cognitive activity is always about making inferences concerning the hidden causes of sensory phenomena. Advocates of the BBH hold that the processes used in mindreading are the same basic kind used elsewhere in every variety of cognition – including acts of basic perception. The only difference in the case of social cognition is the target of the activity.

Thus, just as non-social objects in our environment are causes of our visual input, the mental states of other people are a part of the physical structure of the world that produces the stream of sensory impressions that our brains receive. In this view, mentalising occurs implicitly and shares a fundamental similarity with the representation of non-social objects: each is a natural result of the brain's endeavour to best explain its sensory input (Palmer, Seth, & Hohwy, 2015, p. 377).

There are no simple one-to-one links between sensory experience and its possible causes, which can be many and various. Things are even more tricky in the cases of the mental states that are presumed to lie behind and cause behaviour. Unlike the causes of sensory simulation that lie at the shallow end of the perceptual pool, mental states are much more hidden, usually staying at the deep end.

By engaging with the world to test hypotheses, over and over again, cognisers actualise and maximise their learning potential, gradually improving their accuracy in representing the causal structure of the world. As Hohwy (2018) puts it, 'the ability to minimise the prediction error over time depends on building better and better representations of the causes of sensory inputs. This is encapsulated in the very notion of model revision in Bayesian inference' (p. 134).

An efficient system is one that 'knows' how to determine what is relevant within a context. Social contexts are much more complex than simple sensory feature detection. Being appropriately sensitive to varying contexts, according to the BBH, is a matter of being able to flexibly adjust the degree of attention given to particular sensory inputs (Hohwy & Palmer, 2014; Van de Cruys, Perrykkad, & Hohwy, 2014). This process, known as precision weighting, is effectively the capacity for determining the relevance of sensory inputs, differentiating between noise and signal. Clark (2016) describes it in terms of the system's ability to 'to treat more or less of the incoming sensory information as 'news', and more generally in the ability to flexibly to modify [sic] the balance between top-down and bottom-up information at various stages of processing ...' (p. 226).

To be effective, the precision-weighting of inferences has to be context-sensitive (Ward, 2018; Van de Cruys et al., 2019). For well-adapted systems, learning and experience is the means by which they come to determine the relevance of evidence in the form of sensory inputs by asking whether these contain content that contradicts and should thus revise what is known or expected. Over time, the system becomes better and better at making these adjustments through a bootstrapping process, learning 'from [changing] regularities in the sensory input' (Van de Cruys et al., 2019, p. 165). This is a form of 'experience-dependent learning that accompanies evidence accumulation' (Friston, 2018, p. 5; see also Bzdok and Ioannidis, 2019). The outcome of such learning is that the system gains the ability to attend to what is relevant and ignore what is not.

As discussed in the previous section, being able to cope with context-sensitivity is particularly important in social situations considering that they 'always vary in their sensory details' (Van de Cruys et al., 2019, p. 165). Crucially, it should now be clear why the Bayesian Brain TT proposals about what underlies social cognition look more promising than their Mental Module TT rivals. This is because according to the BBH the brain is always attempting to calculate which state of the world is '*most likely* to be causing the sensory input that our brain receives, given prior beliefs about these causes that are furnished by previous experiences, development, and evolution' (Palmer et al., 2015, p. 377).

The BBH differs from a ToMM precisely in not being encapsulated. The BBH assumes that there is open channel communication – back and forth – that allows inferences to be updated, such that the whole predictive effort is informed by and updated at all levels. The brain's hypotheses and inferences about expected causes of the behaviour of others are thus:

> situated as part of a causal hierarchy, and share reciprocal interactions with higher and lower levels of representation. Thus, mental state inferences are statistically constrained by representations of longer-term expectations – perhaps regarding, for example, the kind of mental states that people tend to have in a given context, or the sense of your friend's mood that has been reflected in a variety of her behaviours since she showed up to the restaurant, or even culturally defined social contexts and norms.
>
> (Palmer et al., 2015, p. 378)

Not only does the BBH overcome the problem of how to account for the context-sensitive use of a ToM in situ, it has been claimed that the approach holds out the promise of illuminating 'a variety of pathologies and disturbances, ranging from schizophrenia and autism to "functional motor syndromes"' (Clark, 2016, p. 3).

Palmer et al. (2015) offer a specific BBH proposal about what explains the signature features of the social cognitive profiles of autistic individuals. Crucially, these researchers conjecture that the autistic social cognitive profiles are a result of the autistic brain's failure to generate relevant inferences and thus to update or revise expectations about other minds. Autism, on this theory, is the result of a deficient precision estimation system.

Autistic individuals lack the flexibility to determine what is relevant because their brains treat too many sensory inputs as signals. This would explain autistic hypersensitivity or sensory overload (Clark, 2016; Pellicano & Burr, 2012) and autistic hypo-responsivity, which is the absence of context-sensitive responses (Van de Cruys et al., 2019). Accordingly, Clark (2016) suggests, this 'would result in a constant barrage of information demanding further processing and might plausibly engender severe emotional costs' (Clark, 2016, p. 225).

Developing this proposal, Clark (2016) suggest that 'autistic subjects can construct and deploy strong priors but may have difficulties applying them' (p. 226). Yet impaired precision-weighting capacities may:

> lead to overspecific, overfitted internal models that will less efficiently explain away sensory inputs of, for example, social situations that always

vary in their sensory details. The sparse, generalizable hidden causes that explain inputs best are not formed (or properly applied)

(Van de Cruys et al., 2019, p. 165)

On this view, incapable of sifting out what is and is not relevant, the autistic brain may be unable to form or update its generative model to get an accurate representation of what causes and lies behind the behaviour of others. If the brains of autistic individuals consistently fail to advance appropriate inferences then they will not have the opportunity to improve the accuracy of their representations of the mental causes that allegedly drive the behaviour of others.

The animating idea behind the BBH is that the work of brains is analogous to what scientists do when making inferences in an effort to best explain phenomena. Brains advance hypotheses, which are developed, refined and improved as those theories are tested against what the world has to offer. As Hohwy (2019) presents it, the BBH's core commitment is that 'perceptual inference is a process that arrives at revised models of the world, which accurately represent the world' (p. 166).

In an earlier publication, Hohwy draws attention to the longstanding analogy, drawn by both Helmholtz and Gregory, 'between perception and scientific hypothesis testing' (Gregory, 1980; Helmholtz, 1860/1962; Hohwy, 2013, p. 77). Clark (2016) makes a direct comparison between what scientific experimenters and brains do (p. 95). Likewise, Yon, de Lange, & Press (2019) note that BBH talk of perceptual inference 'likens perceptual processing to the scientific process' (p. 6).

How seriously should we take the brain–scientist analogy? In what respects should we take it seriously? Sometimes, the analogy is described as 'merely a heuristic description' (Hohwy, 2019, p. 166). Yet, to let go of a realistic reading of the idea that the brain poses and updates inferences raises deep questions about the status of the explanations offered by the BBH. As Hohwy (2018) observes:

> If the inferential aspect is not kept in focus, then it would appear to be a coincidence, or somehow an optional aspect of perceptual and cognitive processes that conform to what Bayes's rule dictate. Put, differently, anyone who subscribes to the notion of predictive processing must also accept the inferential aspect. If it is thrown out, then the 'prediction error minimization' part becomes a meaningless, unconstrained notion.
>
> (p. 132)

In line with the claim that the BBH embeds a non-negotiable commitment to the brain trading in inferences, Hohwy (2018) tells us that the BBH operates with 'a concrete sense of "inference" where Bayes's rule is used to update internal models of the causes of the input in the light of new evidence' (p. 131).

Hohwy also tells us that the Bayesian way in which the brain updates its models and theories is unlike what goes on in explicit deductive reasoning in key respects. Hence, the BBH does not entail that the brain actually gets its work done by an 'overly intellectual application of theorems of probability theory' (Hohwy, 2018, p. 132). As such, the way the BBH construes the brain's inferences differs 'from the use of the term "inference" to describe a higher-order, cognitive and consciously effortful process' (Palmer et al., 2015, p. 379).

This leads us to the view that the brain's inferences are implicit and unconscious, unlike those inferences of scientists that are explicit and conscious. The brain's inferences are swifter and abductive in character and thus unlike inferential operations of the sort found in deductive proofs. Yet, for all that, the brain's inferences are like the inferences of scientists in being contentful and aiming to get an accurate depiction of the true causal structure of reality. That is what all inferences have in common. That is why 'mentalising slots into predictive processing as constituting the same kind of unconscious inference that the brain is already engaged into represent [sic] its environment' (Palmer et al., 2015, p. 378; see Hohwy & Palmer, 2014; Kilner, Friston, & Frith, 2007).

What we can conclude from this is that the BBH, like all versions of TT, is committed to the assumption that we always and everywhere understand others by advancing and improving inferences about the hidden causes of their observable behaviour. This TT picture of what underpins social cognition – and what explains typical and atypical varieties – trades upon and gets its life from the assumption that everyday acts of social cognition are rightly characterised in terms of mindreading. It is that assumption that suggests that we must always adopt a spectatorial stance toward others, even if, unbeknownst to us. In the concluding section, we expose how the spectatorial assumption mischaracterises everyday social cognition and why our exposé should cast doubt on the BBH's attempted explanations of the social cognitive profiles of autistic individuals.

Characterising social cognition correctly: diversity not deficit

TT explanations of everyday social cognition – both of the general population as well as that of autistic individuals – are only attractive to those who adhere to a mindreading picture of such cognition. The mindreading picture is bound up with a host of metaphors. Under its sway, philosophers are wont to claim that individuals have no direct access to other people's minds; that mental states are the out-of-sight, hidden causes that drive behaviour; that in trying to understand what drives another's behaviour we need to posit hypothetical entities in our efforts to accurately get at hidden causes, and so on and on.

The mindreading characterisation of social cognition gets its life from the spectatorial assumption that holds that our situation with respect to others is fundamentally that of a scientific spectator to target phenomena (Hutto, 2004). That assumption is fostered by thinking that the primary point and pervasive purpose of everyday social cognition is to bridge an assumed epistemic gap that exists between us and others for the purposes of accurately depicting the mental states that move them.

Despite the mindreading picture's status as the received view, many philosophers have argued that, on close inspection, modelling everyday social cognition on a scientific enterprise paints a distorting picture of its character (Hutto, 2004, 2008; McGeer, 2007, Ratcliffe, 2007). Positively characterised, our everyday social cognition is bound up with engaging with the attitudes and emotions of others, understanding their projects and commitments, trusting or not trusting the accounts that give us why they do what they do. In these practices we are not taking up a scientific stance towards others.

The point of this reminder is not, pace Carruthers, to say that sub-personal theorising cannot be interactive because it is third-personal. The BBH demonstrates that scientific theorising can 'straddle the interaction–observation dichotomy' (Schönherr & Westra 2017, p. 5). The objection to the TT framework made by its so-called phenomenological critics is more fundamental: it is that our everyday engagements with one another are misdescribed when they are depicted as being essentially theoretical in character. We are not always and everywhere attempting to discover the underlying causes of another's behaviour. This is because we are interested in the other's reasons for acting and the best way to get at those reasons is to be told what they are without even having to ask. To understand those social exchanges aright is, to use McGeer's (2007) words, to recognise that we do not 'interact with one another as scientist to object, as observer to observed' (p. 146).

We maintain that reasons to doubt that the mindreading picture paints a reliable portrait of everyday social cognition are also reasons to doubt that TT proposals, of whatever stripe, can provide the best explanations of social cognition. Thus, relinquishing the spectatorial assumption raises questions about the explanatory appropriateness of applying a TT gloss to characterise so-called sub-personal processes that underwrite social cognition. In other words, giving close attention to the character of our everyday social cognitive practices should make us wary of taking the brain–scientist analogy at all seriously. Rejecting the mindreading answer to the characterisation challenge should make us question the credibility that scientific inference, or even something near enough, really lies at the heart of all of our social cognitive endeavours – namely, that scientific inference really is the driving force in the engine of social cognition.

Some have denied that this sort of conclusion follows. They hold that phenomenological reflections on the character of everyday social cognition do not strongly constrain theorising about its sub-personal drivers. In this vein Spaulding (2018) argues that careful introspection of what it is like to engage in social cognition should not constrain theorising about its underlying mechanisms because 'many of our social interactions consist in tacit or implicit mindreading, i.e. subconsciously explaining and predicting targets' behavior on the basis of attributed mental states' (pp. 14–15).

Apart from begging the question at issue, the trouble with this line of defence is that phenomenological critiques of the mindreading depiction of everyday social cognition are not based, pace Spaulding, on 'careful introspection' of our phenomenology. Rather they are based on giving careful attention to the character of our everyday practices (Hutto, 2013; Hutto & Satne, 2018). In addition, there have been other substantial critiques that raise doubts about the tenability of accounting for the source and basis of the implicit, unconscious contentful inferences upon which the mindreading story and TT explanations rely (see, e.g. Hutto & Myin 2013, 2017). Together, this clutch of objections constitutes a pincer movement that brings the mindreading–TT package into question from two directions, above and below.

Under pressure, Spaulding (2018) admits that when characterising the subpersonal processes that allegedly underwrite social cognition 'one could substitute "interpretation" and "anticipation" for explanation and prediction' (p. 15). That is certainly closer to the mark and, if we are right, that move has much better prospects of bringing attempts at providing sub-personal explanations of social cognition into line with its actual character.

In this light, Spaulding's proposed adjustment to the TT gloss is a step in the right direction. An even bigger and better step, in our view, would be to embrace an enactive account of cognition and abandon the quest to find underlying subpersonal mechanisms that explain cognitive phenomena all together (see Hutto & Myin, 2013, 2017).

It matters which of these philosophical frameworks we adopt for thinking about cognition – the choice has practical and ethical significance. Consider that, on the one hand, it is easy to espouse that autistic individuals deserve our full respect and support as 'persons who try to make sense of themselves and the world' (Procter, 2001, p. 117). Yet, on the other hand, it is equally easy to hold – at the same time – that the impaired mindreading of autistic individuals bars them from making adequate choices in the social domain and, as such, 'caring for them may require making these choices for them' (Procter, 2001, p. 114). This way of reasoning can lead to bad outcomes, as in Melanie Yergeau's case. She reports a harrowing story in which she was forcibly detained by therapists-cum-faculty in which, in her words: 'I found myself deeper within a narrative of neurological determinism ... Regardless of what I said, it was my autism saying it' (Yergeau, 2013). Enactivist approaches to mind and cognition give us tools for resisting rather than encouraging the idea that who 'we' are reduces to something inside us, that who 'we' are is the product of something inside us – the intelligent, subpersonal activity of our brains.

The foregoing analysis is not designed to deny or obscure the fact that certain autistic and cognitively typical individuals, given their particular cognitive capacities and profiles, find some kinds of social cognitive tasks difficult, perhaps even impossibly so. It does, however, serve to remind us that the cognitive challenges and achievements in question are challenges and achievement of individuals – of persons – and that these are not best explained by focusing solely or primarily on sub-personal parts of people.

Had there been more space, it would have been illuminating to provide more detail about enactivist alternatives to the mainstream cognitivist approaches that take the mindreading characterisation of social cognition as given. In lieu of doing so in more depth, we must be satisfied with dwelling on this lesson for now: once we stop thinking of the main action of social cognition as happening in the heads of individuals and put it back in the space of interactions themselves it becomes clear that successful social engagement is a joint responsibility.[2] It is best conceived of as a shared endeavour in which adjustments need to be made by all parties involved to ensure successful outcomes. Success in these tasks, given their true point and purpose, is not something that can be achieved privately and separately in the heads or brains of the individuals involved, however remarkable those individuals may be at inferring each other's mental states.

Enactivists conceive of cognition in terms of dynamic, 'out-of-the-head', world-involving activities. Emphasising these aspects, they make much of the metaphor that cognitive engagements are a matter of 'laying down a path in walking'. That metaphor helps to illuminate that in all cases, when it comes to completing cognitive tasks there are other possible ways of getting to the same place. Getting there by other ways might require going slower, or taking a different path than the beaten one, or they might require the cooperation and assistance in meeting one part of the way.

The pivotal point is that we have reason to surrender the idea, enshrined in the intellectual individualism embraced by mainstream cognitivism, that successful social cognition depends on and aims at a gap-bridging epistemic achievement. Concomitantly, we have reason to avoid the idea that such gap-bridging can only be achieved by the theorising that goes on in the brains of individuals.[3] If we manage to resist these prevalent pictures, we open the door for thinking of the success of social cognitive tasks in ways that are not purely epistemic and to recognising that the success of such engagements depends on and is the mutual responsibility of all of the individuals, both autistic and cognitively typical, involved in social encounters.

Research for this article was supported by the Australian Research Council Discovery Project 'Mind in Skilled Performance' (DP170102987).

Notes

1 In this chapter we focus exclusively on TT proposals: however, our conclusions generalise. They apply with equal force to any Simulation Theory, or ST, proposal which assumes that neutrally based mindreading involves some kind of inference. ST proposals hold that inferences about other minds are achieved by co-opting or reusing planning or practical reasoning systems – rather than forming theories about goal planning, as is the wont of TT proposals (see, e.g. Gordon, forthcoming). However, on our analysis, the difference between ST and TT about the character of the central processes involved in modelling other minds – while technically interesting – is unimportant with respect to the larger concerns we raise in this paper. This is because both classes of theory, TT and ST, characterise everyday social cognition in mindreading terms. Our objections apply to any explanation that takes the mindreading characterisation of everyday social cognition for granted. Hence, our analysis applies with equal strength to any mentalising proposal – whether ST or T – which assumes that it is the job of the brain to infer

the goals or reasons (or otherwise 'the causes') that lie behind another agent's observed behaviour.

2 The double empathy hypothesis (Milton, 2012) states that cognitively typical people have just as much difficulty empathising with autistic people as vice versa because difficulties will inevitably arise when different cognitive styles are in communication. This hypothesis is based on the view that autism involves autistic cognition, not impaired neurotypical cognition.

3 We do not discount the importance of brain-based aspects of cognition. The monotropism theory (Murray, Lesser, & Lawson, 2005) is regarded by many autistic scholars, including the third author, as capable of describing the features of autism, including differences in sensory perception and sociality.

References

American Psychiatric Association (2013). *Diagnostic and statistical manual of mental disorders* (5th ed.). Washington, DC: American Psychiatric Publishing.

Baron-Cohen, S. (1995). *Mindblindness: An essay on autism and theory of mind.* Cambridge, MA: MIT Press.

Baron-Cohen. (2019). The concept of neurodiversity is dividing the autism community. *Scientific American*, April 30.

Baron-Cohen, S. (2000). Theory of mind and autism: A fifteen-year review. In S. Baron-Cohen, H. Tager-Flusberg, & D. J. Cohen (Eds.), *Understanding other minds: Perspectives from developmental cognitive neuroscience* (pp. 3–20). Oxford: Oxford University Press.

Belmonte, M. (2009). What's the story behind 'theory of mind' and autism? *Journal of Consciousness Studies, 16*(6–7), 118–139.

Brewer, N., Young, R. L., & Barnett, E. (2017). Measuring theory of mind in adults with autism spectrum disorder. *Journal of Autism and Developmental Disorders, 47*(7), 1927–1941.

Brüne, M. (2005). 'Theory of mind' in schizophrenia: a review of the literature. *Schizophrenia bulletin, 31*(1), 21–42.

Bzdok, D., & Ioannidis, J.P. (2019). Exploration, inference, and prediction in neuroscience and biomedicine. *Trends in Neurosciences, 42*(4), 251–262.

Carruthers, P. (2011). *The opacity of mind: An integrative theory of self-knowledge.* Oxford: Oxford University Pres.

Clark, A. (2016). *Surfing uncertainty: Prediction, action, and the embodied mind.* Oxford: Oxford University Press.

Fodor, J. A. (1983). *The modularity of mind.* Cambridge, MA: MIT Press.

Friston, K. (2018). Does predictive coding have a future? *Nature Neuroscience, 21*(8), 1019–1021.

Gilleen, J., Xie, F., & Strelchuk, D. (2017). Distinct theory of mind deficit profiles in schizophrenia and autism: A meta-analysis of published research. *Schizophrenia Bulletin, 43*(Suppl 1), S22.

Gordon, R. (forthcoming). Simulation, predictive coding, and the shared world. In K. Ochsner & M. Gilead (Eds.), *The neural bases of mentalizing.* New York: Springer.

Gregory, R. L. (1980). Perceptions as hypotheses. *Philosophical Transactions of the Royal Society of London. B, Biological Sciences, 290*(1038), 181–197.

Heyes, C. (2018). *Cognitive gadgets.* Cambridge, MA: Harvard UP.

Hohwy, J. (2013). *The predictive mind.* Oxford: Oxford University Press.

Hohwy, J. (2018). The predictive processing hypothesis. In *The Oxford handbook of 4E cognition* (pp. 129–145). Oxford: Oxford University Press.

Hohwy, J. (2019). Prediction error minimization in the brain. In *The Routledge handbook of the computational mind* (pp. 159–172). London: Routledge.

Hohwy, J., & Palmer, C. (2014). Social cognition as causal inference: Implications for common knowledge and autism. In M. Gallotti & J. Michael (Eds.), *Perspectives on social ontology and social cognition* (pp. 167–189). Dordrecht: Springer.

Hutto, D. D. (2004). The limits of spectatorial folk psychology. *Mind & Language, 19*(5), 548–573.

Hutto, D. D. (2008). *Folk psychological narratives: The sociocultural basis of understanding reasons*. Cambridge, MA: MIT Press.

Hutto, D. D. (2013). Enactivism, from a Wittgensteinian point of view. *American Philosophical Quarterly, 50*(3), 281–302.

Hutto, D. D., & Myin, E. (2013). *Radicalizing enactivism: Basic minds without content*. Cambridge, MA: MIT Press.

Hutto, D. D., & Myin, E. (2017). *Evolving enactivism: Basic minds meet content*. Cambridge, MA: MIT Press.

Hutto, D. D., & Satne, G. (2018). Naturalism in the Goldilocks zone: Wittgenstein's delicate balancing act. In K. M. Cahill & T. Raleigh (Eds.), *Wittgenstein and naturalism* (pp. 56–76). London: Routledge.

Karmiloff-Smith, A., Klima, E., Bellugi, U., Grant, J., & Baron-Cohen, S. (1995). Is there a social module? Language, face processing, and theory of mind in individuals with Williams syndrome. *Journal of cognitive Neuroscience, 7*(2), 196–208.

Kilner, J. M., Friston, K. J., & Frith, C. D. (2007). Predictive coding: an account of the mirror neuron system. *Cognitive processing, 8*(3), 159–166.

Livingston, L. A., Colvert, E., Social Relationships Study Team, Bolton, P., & Happé, F. (2019). Good social skills despite poor theory of mind: exploring compensation in autism spectrum disorder. *Journal of Child Psychology and Psychiatry, 60*(1), 102–110.

Maibom, H. (2009). In defence of (model) theory theory. *Journal of Consciousness Studies, 16*(6–7), 360–378.

Matthews, L. J. (2019). Isolability as the unifying feature of modularity. *Biology & Philosophy, 34*(2), 20.

McGeer, V. (2007). The regulative dimension of folk psychology. In D. D. Hutto & M. M. Ratcliffe, (Eds.) *Folk psychology re-assessed* (pp. 137–156). Springer: Dordrecht.

Milton, D. (2012). So what exactly is autism? Autism Education Trust.

Murray, D., Lesser, M., & Lawson, W. (2005). Attention, monotropism and the diagnostic criteria for autism. *Autism, 9*(2), 139–156.

Palmer, C. J., Seth, A. K., & Hohwy, J. (2015). The felt presence of other minds: Predictive processing, counterfactual predictions, and mentalising in autism. *Consciousness and Cognition, 36*, 376–389.

Pellicano, E., & Burr, D. (2012). When the world becomes 'too real': a Bayesian explanation of autistic perception. *Trends in Cognitive Sciences, 16*(10), 504–510.

Pezzulo, G. (2017). Tracing the roots of cognition in predictive processing. In T. Metzinger & W. Wiese (Eds.), *Philosophy and predictive processing*. Frankfurt am Main: MIND Group.

Procter, H. G. (2001). Personal construct psychology and autism. *Journal of Constructivist Psychology, 14*(2), 107–126.

Quilty-Dunn, J. (2019). Attention and encapsulation. *Mind & Language*. https://doi.org/10.1111/mila.12242

Raftopoulos, A. (2019). *Cognitive penetrability and the epistemic role of perception.* Basingstoke: Palgrave Macmillan.

Ratcliffe, M. (2007). From folk psychology to common sense. In D. D. Hutto & M. M. Ratcliffe, (Eds.), *Folk psychology re-assessed* (pp. 223–243). Dordrecht: Springer.

Scholl, B. J., & Leslie, A. M. (1999). Modularity, development and 'theory of mind'. *Mind & Language, 14*(1), 131–153.

Schönherr, J., & Westra, E. (2019). Beyond 'interaction': How to understand social effects on social cognition. *The British Journal for the Philosophy of Science, 70*(1), 27–52.

Spaulding, S. (2010). Embodied cognition and mindreading. *Mind & Language, 25*(1), 119–140.

Spaulding, S. (2018). *How we understand others: Philosophy and cognition.* Abingdon: Routledge.

Ure, A., Rose, V., Bernie, C., & Williams, K. (2018). Autism: One or many spectrums? *Journal of Paediatrics and Child Health, 54*(10), 1068–1072.

Van de Cruys, S., Perrykkad, K., & Hohwy, J. (2019). Explaining hyper-sensitivity and hypo-responsivity in autism with a common predictive coding-based mechanism. *Cognitive neuroscience, 10*(3), 164–166.

von Helmholtz, H. (1962). Handbuch der physiologischen optik. 1860/1962. (trans by J. P. C. Southall). London: Dover.

Ward, J. (2018). Individual differences in sensory sensitivity: A synthesizing framework and evidence from normal variation and developmental conditions. *Cognitive Neuroscience, 1–19.*

Yergeau, M. (2013). Clinically significant disturbance: On theorists who theorize theory of mind. *Disability Studies Quarterly, 33*(4).

Yon, D., de Lange, F. P., & Press, C. (2019). The predictive brain as a stubborn scientist. *Trends in Cognitive Sciences, 23*(1), 6–8.

Part VI

Moving forwards

Chapter 13

Neuronormativity in theorising agency

An argument for a critical neurodiversity approach

Dieuwertje Dyi Huijg

'Agency' is a very diversely employed concept. In sociology it is grounded in the 'agency-structure debate', but more generally it revolves around questions of freedom and determinism, including of (free) will and self-determination. The discussion boils down to the question whether we can – always or ever – 'determine' our own actions. Not only does this raise the question who or what constrains, coerces, limits, or even precludes agency, but also who does and who does not 'have' agency to start with. Problematically, some argue that (only) all those with a *normal* 'state of mind' or 'mental ability' would have agency (e.g. Archer, 2016 [2007], p. 167; Barnes, 2000, pp. 10–11, 107–108). This raises the question what precisely the empirical and theoretical relation of this supposedly 'normal state of mind or mental ability' is to (the conceptualisation of) agency. At this stage it is useful to consider, for example, that theorisations of agency are generally centred around ideas and ideals of 'rationality' and 'rational action' (e.g. Campbell, 1996, p. 6; Hoggett, 2001; Stout, 2005) and how research on neurodivergence and agency can be centred around ideas and ideals of 'morality' and 'moral agency' (e.g. Kennett, 2002; Vehmas, 2011). Reflecting in this sense about agency is not merely an intellectual endeavour; the assessment whether or not an individual can be considered 'agential' and 'agentic'[1] – that is, respectively being capable of mobilising agency and acting upon this – is not just an existential or theoretical matter, but arguably forms the foundation for legal, police, health, and educational thought, policies, and practices. In other words, a battle about the conceptualisation of agency is relevant because of its real, explicit and implicit, implications.

While there is no agreement on the conceptualisation of agency, many scholars agree with a working definition of agency, stripped to its conceptual bare minimum, as the 'capacity to act' (e.g. see Ahearn, 2001, p. 112 in Campbell, 2009, p. 408; Sibeon, 1999, p. 139) – or, as I have argued elsewhere, 'the capacity to act or to not act' (Huijg, 2012, 2020). An action,[2] as the result of mobilising this capacity, can refer to basic and complex doings; from raising one's arm, raising one's arm to fire a gun, up to negotiating a gun deal between two countries.[3] Feminist scholars particularly have offered important contributions about agency; for instance, the question whether we should talk about *victims* or about *survivors* of domestic violence demanded a response to the question whether experiencing abuse precludes agency or not. Being at the receiving end of a gun or other forms of violence negatively impacts, empirically and theoretically, one's potential for agency due to conditions of unfreedom (i.e. coercion or worse), but agency is not in and of itself restricted to empowerment, nor to one particular ethical or ideological quality. At the same time, these feminist considerations required a re-evaluation of the role of gender and hegemonic masculinity specifically in understanding agency. As agency is not the sole property of (male) people in power, nor limited to exerting violence, neither is agency a capacity of neurotypical people only. Before addressing this, I briefly want to turn to 'agency' itself.

What is central in the understanding of agency as the individual's 'capacity to act' is that this individual needs to *cause* that particular action, with the availability of an option to act differently, and that these actions are not caused by coercion or otherwise (e.g. by means of a gun). What sometimes gets overlooked is that a discussion of agency requires an understanding of 'capacity'. Dictionary understandings offer that (a) capacity is 'the ability or power to do or understand something' (Oxford Dictionaries, 2016b). In turn, (an) ability is 'the possession of the means or skill to do something' (Oxford Dictionaries, 2016a). By introducing the interconnectedness of 'capacity' and 'ability', I point to a link between 'agency' and 'ability'. This introduces disability and, therein, neurodiversity.[4]

The aforementioned working definition of agency as the capacity to act refers to the ability *to act* as well as to the *ability* to act. In that light, perhaps the dis-capacity to act, as one could think of an individual *not* 'having' agency, can also be made sense of as dis-ability. Questions around how the functioning of the body impacts the individual's ability to act, then, invite 'dis-ability' as well as 'disability' into the conversation about agency. However, the disabled body's functioning is also used, presumably by non-disabled scholars, to point to the conceptual boundaries of agency. To exemplify such functionings and boundaries, Stout (2005, p. 5) explains that an individual with Tourettes' 'involuntary tics, both physical and verbal,' are not 'intentional actions'. Since they '[just] happen, like blinking', the individual, is the argument, does not *cause* and, hence, does not *do* the(ir) ticking and the tics do not represent their 'intending' (pp. 5, 7); rather than grounded in intentionality, tics are caused neurologically. As Tourettes (or, for that matter, anyone's) tics are marked by a dis-ability to *not* 'do something', namely ticking, tics are actions that fall outside the realm of agency. While pointing to the non-agentic character of Tourettes tics might be valid and useful – after all, this could form the ground for financial and personal state support – the relationship between agency and neurodiversity requires more critical attention.

Similar to race, sexuality, and gender, I employ neurodiversity[5] here as a(n) analytical) social category that refers to, on the one hand, 'neurodivergent' people – e.g. people with Tourettes, ADHDers, autistics – and, on the other hand, 'neurotypical' people. To emphasise, the 'neuro' here refers to a *social categorical* 'neuro-difference', which neither alludes to an affirmation of a biomedical qualification, nor to a neutral social difference. The understanding of neurodiversity as a category of power relations and social hierarchisation becomes clear in Graby's (2015, p. 235) discussion of neurodiversity in terms of 'minority neurotypes' and 'majority neurotypes.' This interlude is relevant as making sense of disability through a critical neurodiversity lens does not only point to the disabling of neurodivergence, but also to, what I name, the 'neuronormativity' of the theorisation of agency – to which I turn now.

In his argument for the role of inner speech in agency, Wiley (2010, p. 25) leans on a particularly, what might be called, 'neuro-ableist' approach; he maintains that 'people who have little or no ability to engage in inner speech' – such as 'ADHD children[6] (Barkley, 1997, pp. 278–282) [and] most autistics (Whitehouse, 2006)' – 'also seem to have little foresight into the consequences of their actions.' Where Stout emphasised 'intentionality', Wiley argues that *any* capacity for agency is grounded in the presence, or perhaps mobilisation, of 'inner speech' and 'foresight'.[7] Consequently, Wiley (2010, p. 25) claims, 'if inner speech is absent or impaired, people have a weakened power of agency'. In other words, not only is the ability *to (not) act* and the *ability* to (not) act impacted, autistics' and ADHDers' capacity for agency *itself* is impaired – if not dis-abled. Indeed, Wiley suggests that a neurodivergent mind precludes a (strong) 'power of agency'. Postponing the question whether 'inner speech' (etc.) and 'foresight' are conceptual conditions for agency, the idea that autistics and ADHDers would, as a class, lack agency or be weakly agential is peculiar.

Presumably the reader will agree that academic writing and editing does require agency and numerous simple and complex agentically mobilised actions. If we can agree on that, then hopefully the irony is not lost on the reader that various of the chapters in this volume are written and edited by neurodivergent scholars. Admittedly, I have not approached the authors and editors with the question whether or not they employed 'inner speech' and 'foresight' while engaging in writing and editing activities. However, each of the authors and editors could have opted for not writing their chapter or editing the volume; there was choice, and since chapters and books do not write and edit themselves, nor does it make sense to suggest that their writing and editing could be 'neurologically caused', deduction suffices here to assess that, one way or another, agency – and presumably not a weak form – was involved in the writing and editing that led to this volume. This nullifies the thesis of neurodivergent folks' dis-capacity – or perhaps de-capacity – for agency.

It might be tempting to follow this claim with the assertion that both neurotypical and neurodivergent people similarly 'have agency' and to end this here. However, as the aforementioned example of ticking suggests, this would be inaccurate; since tics are conceptual markers of Tourettes, someone with Tourettes ticks disproportionately more than someone without, and tics fall outside the realm of agency (cf. Stout, 2005, pp. 5, 7), someone with and someone without Tourettes have a different relation to agency.[8] At the same time, the argument in this chapter *disputes* that acting agentially requires a 'normal state of mind or mental ability' (e.g. Archer, 2016 [2007], p. 167; Barnes, 2000, pp. 10–11, 107–108). Introducing a neurodiversity perspective to this claim suggests, first, that the 'normality' of that mental ability is actually a *neurotypical* ability and, second, that theorisations of agency that take neurotypical minds as the 'normal state' are arguably grounded in *neuronormativity* – if not neuro-ableism. It is important to emphasise that the neurodivergent experience, assessment, and expression of, for instance, time, context, communication with and perception of self and other – thought by others in terms of intentionality, foresight, inner speech etc. – might well impact the autistic, Tourettes, and ADHD mobilisation of agency. Rather than this being ground for the reproduction of a neuronormative conceptualisation of agency, I conclude here by offering a threefold question for further research. First, how does the standard of neurotypicality impact the thinking about agency? Second, how do neurodivergent folk agentically 'do' their actions? And, consequently, last, what can a critical neurodiversity lens contribute to the conceptualisation of agency from a neurodivergent standpoint?

Notes

1 I discuss this in my doctoral thesis on 'intersectional agency' (Huijg, 2020).
2 For a discussion on the conceptualisation of 'action', and on the role of 'social' therein, see Huijg (2020).
3 Note that 'killing someone by gun' cannot necessarily be considered an agentic action, as it conflates agency, action, and the consequences of action (e.g. Barnes, 2000, pp. 9–10, 18).
4 I understand neurodivergence as a form of disability – in terms of dis/ability as a social category rather than as impairment.
5 The autistic activist Singer coined the term 'neurodiversity' in 1998 (see Singer, 2016, p. 9) and it has been in development since (e.g. see this volume).
6 The question whether children can be fully capable of agency in the first place is a related but separate discussion.
7 Different theorists, schools of thought, and disciplines emphasise different conditions. For instance, I discuss choice, consciousness, reflexivity, and orientatedness as conditions (Huijg, 2020).
8 The argument here is certainly not that those with Tourettes have no or a 'weakened power' of agency.

References

Archer, M. S. (2016 [2007]). Reflexivity as the unacknowledged condition of social life. In: Tom Brock, Mark Carrigan & Graham Scambler (Eds.), *Structure, culture and agency: Selected papers of Margaret Archer* (pp. 165–183). Abingdon: Routledge.
Barnes, B. (2000). *Understanding agency: Social theory and responsible action*. London: Sage.
Campbell, C. (1996). *The myth of social action*. Cambridge: Cambridge University Press.
Campbell, C. (2009). Distinguishing the power of agency from agentic power: A note on Weber and the 'black box' of personal agency. *Sociological Theory, 27*(4), 407–418.
Graby, S. (2015). Neurodiversity: Bridging the gap between the disabled people's movement and the mental health system survivors' movement? In Helen Spandler,

Jill Anderson & Bob Sapey (Eds.), *Madness, distress and the politics of disablement* (pp. 231–244). Bristol: Policy Press.

Hoggett, P. (2001). 'Agency, rationality and social policy'. *Journal of social policy*, 30(01), 37–56.

Huijg, D. D. (2012). Tension in intersectional agency: A theoretical discussion of the interior conflict of white, feminist activists' intersectional location. *Journal of International Women's Studies*, *13*(2), 3–18.

Huijg, D. D. (2020). *Intersectional agency: A theoretical exploration of agency at the junction of social categories and power, based on conversations with racially privileged feminist activists from São Paulo, Brazil.* (Unpublished doctoral dissertation). University of Manchester, Manchester. Available at www.research.manchester.ac.uk/portal/en/theses/intersectional-agency-a-theoretical-exploration-of-agency-at-the-junction-of-social-categories-and-power-based-on-conversations-with-racially-privileged-feminist-activists-from-sao-paulo-brazil(a2dae487-1146-4fd1-b451-5b05a76b77fc).html

Kennett, J. (2002). Autism, empathy and moral agency. *The Philosophical Quarterly*, *52*(208), 340–357.

Oxford Dictionaries. (2016a). Ability. *Oxford dictionaries* [online]. Retrieved from www.oxforddictionaries.com/definition/english/ability.

Oxford Dictionaries. (2016b). Capacity. *Oxford dictionaries* [online]. Retrieved from www.oxforddictionaries.com/definition/english/capacity.

Sibeon, Roger (1999). Agency, structure, and social chance as cross-disciplinary concepts'. *Politics*, *19*(3), 139–144.

Singer, J. (2016). *NeuroDiversity: The birth of an idea.* Amazon Digital Services.

Stout, R. (2005). *Action.* Chesham: Acumen.

Vehmas, S. (2011). Disability and moral responsibility. *Trames*, *15*(2), 156–167.

Wiley, N. (2010). Inner speech and agency In Margaret Archer (Ed.), *Conversations about reflexivity* (pp. 17–38). Abingdon: Routledge.

Chapter 14

Defining neurodiversity for research and practice

Robert Chapman

Neurodiversity means a lot of different things to different people. For Singer (1999) and Blume (1998), it was more associated with an 'ecological society' where minority minds are valued in light of, and helped to find, their niche. By contrast, Walker's (2014) influential definition distinguishes between the fact of neurological diversity (a manifestation of genetic diversity), and the neurodiversity paradigm, which is more about depathologising and instead politicising neurodivergence. I myself have analysed it both as a political idea, associated with social models of disability (e.g. Chapman, 2019) and as a scientific concept, indicating a new way of thinking about function and dysfunction (Chapman, forthcoming). Others have used the term in different ways still.

I will only briefly mention how I think neurodiversity differs from the medical and social models, since that is what I have done in my other chapter (Chapter 4) in this volume. In short, I argue that neurodiversity is anathema to the medical model, but also that there is a technical contradiction between neurodiversity and the social model (or at least the traditional version of it). This regards the concept of 'impairment', which is measured in relation to a species norm in terms of functional ability, which is part of the social model. The issue is that the very idea of neurodiversity seems to me to include a challenge to the reliance on a species norm for assessing (and valuing) our functional abilities at all, in favour of the notion that diversity itself is normal. And if this is the case, then we must find a way to acknowledge differences in functioning in a way that does not rely on the species-norm–based notion of impairment.

As to what neurodiversity means, I will explain why I am ambivalent about definitions. On the one hand, it is important to try and understand, clarify, and analyse neurodiversity, both as a concept and as a movement. In large part it is important to do this because it is a concept that affects many people, and which can be used or abused, in a multitude of ways. Also, how successful the movement is will, to some extent, depend on how viable its underlying concepts and theoretical basis are. Of course, it is also helpful to define terms for more everyday reasons, in so far as we need to understand what others are talking about for successful communication.

Nonetheless, my own understanding has changed considerably, and it continues to do so. In my own case, I have long counted my autism as part of neurodiversity, but I see my post-traumatic stress as a genuine mental disorder. This was initially because I see autism as a natural and valuable manifestation of human genetic diversity (albeit disabled by society), whereas my post-traumatic stress is more a set of unwanted ingrained responses to distressing experiences. In framing it this way, I think I was influenced here by all the talk of 'natural' variations often heard in neurodiversity proponent circles.

But consider a different example. Some research suggests that some cases of personality disorder are, in significant part at least, responses to early traumatic life events. For this reason, I initially assumed that people diagnosed with disordered personalities would, as with my post-traumatic stress, count them as genuine disorders rather than part of natural and valuable neurodivergence. But I have since been convinced that some individuals given those diagnoses also can find the neurodiversity framing helpful and liberating (or at least some do so). And if the neurodiversity framing is as helpful for those labelled as having disordered personalities as it has been for so many autistic people, wouldn't it be better to develop a more inclusive concept of neurodiversity rather than exclude them? And why should it matter if any given set of traits is 'natural' or not anyway? I rather think the focus on whether things are natural or not often detracts from more important goals.

It is because of such considerations that I both think it is vital to critically analyse, but am simultaneously sceptical of attempts to offer final definitions of, neurodiversity. For on the one hand, we do need to be able to distinguish between minority forms of functioning and genuine pathology; but on the other hand, any attempt at definition risks being harmful or exclusionary. With this ambivalence at defining neurodiversity in mind, I will just say two final things.

First, I think that neurodiversity is likely what philosophers call a 'moving target', meaning that the concept will continue to change and 'move' due to complex interactions between those who are categorised by it (including both neurotypicals and neurodivergents), as well as the various relevant institutions it challenges and responds to (psychiatry, education, etc.). In short, it will mean different things at different times. Given this, while I certainly think there are better and worse definitions of neurodiversity, and that it is the kind of idea that can be used or abused, I do not think it is the kind of thing we can or should hope for a final definition of.

Second, though, I can say more about what the concept is *useful* for. Over seven years of working on the subject, I have come to see it being more of an epistemically useful concept than anything else. By 'epistemic' I mean relating to knowledge; and in describing it as being 'epistemically useful', I mean in terms of helping us access and generate new forms of knowledge. From this perspective, a core function of the concept regards how it helps us imagine the world differently to how it currently is. For instance, it helps us to both reimagine pathologised and dehumanised kinds in a more humane and compassionate way and reimagine the world in a way that is less hostile to such kinds. In turn, by adopting a neurodiversity perspecfive, we can alter actual relations; all the way from how we empathise with neurological others on a personal level, to how we design scientific experiments or public spaces. Similarly, within and between neurominorities, it helps us foster not just solidary and resistance, but also grounds the development of shared vocabularies for making sense of our experiences and increasing our understanding of both each other and ourselves. So what starts out first as something epistemically useful, translates into the generation of different social facts, and finally into real world change.

References

Blume, H (1998). Neurodiversity: on the neurological underpinnings of geekdom. *Atlantic*. Available at www.theatlantic.com/magazine/archive/1998/09/neurodiversity/305909/

Chapman, R. (2019). Neurodiversity and its discontents: Autism, schizophrenia, and the social model. In S. Tekin, & R. Bluhm (Ed.), *The Bloomsbury companion to the philosophy of psychiatry* (pp. 371–389). London: Bloomsbury.

Singer, J. (1999). 'Why can't you be normal for once in your life?' From a 'problem with no name' to the emergence of a new category of difference. In M. Corker & S. French (Eds.), *Disability discourse*. Buckingham: Open University Press.

Walker, N. (2014). Neurodiversity: Some basic terms and definitions. Neurocosmopolitanism. Available at: https://neurocosmopolitanism.com/neurodiversity-some-basic-terms-definitions/

Chapter 15

A new alliance?

The Hearing Voices Movement and neurodiversity

Akiko Hart

This chapter is concerned with neurodiversity as a hermeneutic framework, a community, and a politic, to describe, explain, and mobilise the experience of hearing voices. It makes no claim as to the prevalence or otherwise of hearing voices within a specific population, for example on the overlap between the diagnosis of autism and the experience of hearing voices (Milne, Dickinson, & Smith, 2017). Instead it attempts to trace the manifestations of neurodiversity within a distinct culture – the Hearing Voices world.

Hearing voices refers to hearing something in the absence of an external stimulus. While the experience has been commonly framed by twentieth- and twenty-first-century psychology and psychiatry as an auditory verbal hallucination and the symptom of a mental illness such as schizophrenia, in the last three decades the Hearing Voices Movement (HVM) has played an important role in challenging this interpretation (Romme, Escher, Dillon, Corstens, & Morris, 2009). Starting from the position that voices are not necessarily the meaningless symptoms of an illness, but instead a meaningful experience, somehow connected to the life and context of an individual's life, a movement, community, and an identity – that of the voice-hearer – have emerged (Woods, 2013), some of it explicitly rejecting psychiatry, its language, models, and practice (Romme et al., 2009). The Hearing Voices approach is in principle open to a wide variety of ways of understanding voices, and embraces a range of frameworks, from voices as spiritual gifts, to voices as a meaningful response to traumatic life experiences. In practice, however, it is the latter hermeneutic model that dominates and which has been espoused by many individuals within the HVM (Dillon, Bullimore, Lampshire, & Chamberlain, 2012, p. 311).

On the one hand, neurodiversity and the Hearing Voices approach instinctively feel like a good fit, notwithstanding that both are highly diverse, amorphous communities and movements, which also have distinct subcultures as well as regional, political and philosophical variations. Hearing voices as a normal human experience that exists on a continuum implicitly maps on to a neurodiversity framework, where ways of being-in and apprehending the world are considered dimensional, and claims to normativity are challenged. Further, in line with neurodiversity activists, proponents of the Hearing Voices approach position distress as not necessarily the product of the experience or neurotype, but rather as a reaction to discrimination, rejection, and misunderstanding (Graby, 2015). Indeed, mirroring the overlaps between Critical Autism Studies and Mad Studies (McWade, Milton, & Beresford, 2015), some of the political aims of neurodiversity and the HVM coalesce. Particularly salient are the right to self-identify, eschewing a deficit model, celebrating diversity, addressing the legacy of harm caused by the (forced) medicalisation of experiences, and nurturing a community of individuals and mobilising it as a political force as well as a source of inspiration and connection.

However, explicit references to neurodiversity are more sparse within the Hearing Voices literature, despite the Hearing Voices approach being indexed to the wider consumer/survivor/ex-patient (C/S/X) movement, and reflect the contested nature of the field. It is intriguing that there is little that is published, and much of what is captured is in list servs, online groups, and chats, with some conversations making their way to blogs – highlighting perhaps, the multiple marginalisations of these voices, as they struggle to find a home within Disability Studies, mainstream mental health literature, but also the critical mental health world.

Some of this reflects the wider malaise in mental health when it comes to the adoption of the social model of disability, and the complex and oftentimes fraught relationship between the disabled people's movement and the mental health survivor movement (Graby, 2015; McWade et al., 2015; Jones and Kelly, 2015). Then there is the vexed issue of impairment (Graby, 2015; Plumb, 2015) which is vulnerable to a two-pronged attack. On the one hand not all voice-hearers will experience distress or difficulty, and therefore the language of impairment and disablement cannot be used unilaterally to describe voice-hearing as an experience (Sapey, 2011). On the other, the distress around hearing voices that many do experience cannot simply be reduced to discrimination, sanism, or psychiatric oppression, thus upsetting the foundations of the social model of disability (Spandler & Poursanidou, 2019; Plumb, 2015). Within this debate, neurodiversity has been located as a bridge between the disability and survivor moments (Graby, 2015), and seems to have the flexibility to accommodate multiple, competing, and conflicting hermeneutic discourses (Dandelion, 2019) which might enable us to move beyond unhelpful binaries and single narratives.

Yet there are a number of flashpoints between the HVM and neurodiversity that preclude an easy alliance. One is the prevailing view within the former that voices are a meaningful response to trauma or adversity, and that we should be asking 'what has happened to you?' instead of 'what is wrong with you?' (Romme et al., 2009). A narrow trauma-based understanding of voices collides with neurodiversity on a number of levels, not least because neurodiversity is framed by many critical mental health scholars and activists as an essentialist identity category and another iteration of psychiatric diagnosis to be disavowed (Timimi, 2018; Milton & Timimi, 2016). Unlike neurodiversity, trauma-based aetiologies and hermeneutics of voices can also struggle to clearly and consistently reference wider adversity, held back by both their psychological roots and the narratives of recovery that buttress the Hearing Voices Movement, which by privileging individual stories of suffering and healing, can often erase more systemic oppressions (Woods, Hart, & Spandler, 2019).

Where both critics of neurodiversity and critics of the Hearing Voices approach might agree, perhaps, is that by framing the experience as difference and as dimensional, rather than as deficit and categorical, both leave themselves dangerously open to failing those most marginalised: those who need and are entitled to support within the medical-deficit model enacted by welfare and health policies (Recovery in the Bin, 2019; Runswick-Cole, 2014). Another critique applicable to both movements is the extent to which they attend to all forms of marginalisation and exclusion, in particular to the experiences of racialised minorities: is self-defining one's experiences a (white) luxury, when confronted with the multiple material oppressions caused by the policing of racialised bodies and minds?

One intriguing bridge between Hearing Voices and neurodiversity has come through those who understand their voices as multiplicity: as parts, persons, alters, or identities within their body, each with their own history, personality, or story (Cretzinger, 2019; Graby, 2015; The Crisses, 2014; The Redwoods, 2018a, 2018b). Some might identify as having a diagnosis of Dissociative Identity Disorder, and their multiplicity as a consequence of and reaction to severe trauma. However, a growing number use 'multiple' or 'plural' (in opposition to 'singlet') as well as the language and model of neurodiversity to define themselves (The Crisses, 2014; The Redwoods, 2018b). This can generate radically different discussions to the ones being held in parts of the Hearing Voices world, as it troubles the binary of voices as a symptom of a disorder versus voices as a response to trauma. While there can still be hurdles, with, for example, multiplicity struggling to accommodate external voices, multiplicity challenges normative thinking around identity and interiority. By doing so, it can pay attention to other, intersecting forms of marginalisation: witness the troubling of gender within multiplicity activism, which queries not only the idea that our gender aligns with that assigned to us at birth, but that there is a singular and fixed gender identity indexed to us (The Crisses, 2014; The Redwoods, 2018b). This not only exposes the Hearing Voices approach to queer and trans activism and scholarship, but conversely unfurls those worlds to queering madness, distress, and mental illness. In this way, voices understood within the framework of neurodiversity move away from an essentialised identity category (Timimi, 2018), and become instead a political strategy of critique and resistance, echoing dialogues between Mad Studies and Queer Studies (Spandler & Barker, 2016). One expression of this has been the emergence of neuroqueer: being neurodivergent and approaching it as a form of queerness, and a way of thinking and doing that queers not only gender and sexuality but also mental illness (Egner, 2019; Walker, 2015). To neuroqueer as a verb and as an action might address critiques of neurodiversity as identity politics which re-enact a fixed binary politics of 'us and them' (Runswick-Cole, 2014). It might also create space in the Hearing Voices approach for all experiences and understandings, where, in practice, and as with any collective, some experiences can be privileged and others marginalised: 'The Hearing Voices Network is politically problematic for me because it constructs a world of centres and margins – and my own "borderline" madness is positioned as peripheral (Grey, 2016).

A generative alliance with neurodiversity might not only offer room for more understandings and identities, but for doing and thinking differently. In doing so, it might build on the foundations and mission of the HVM as a disruptive political force, by opening it up to (other) material realities of oppression, in order to further affirmative ways of being-in-the-world, as well as liberatory action, policy, and practice. In short, neurodiversity might help bring the Hearing Voices Movement into the twenty-first century.

References

Cretzinger, S. (2019). A brief experience of multiplicity. *Asylum: The Magazine for Democratic Psychiatry* https://asylummagazine.org/2019/06/a-brief-experience-of-multiplicity-by-sharon-cretsinger/ (Accessed 1 November 2019).

Dandelion (2019). Les voix, mes voix et moi. https://medium.com/@sirdandelion/les-voix-mes-voix-et-moi-2fc3b69b8870. (Accessed 1 November 2019).

Dillon, J., Bullimore, P., Lampshire, D., & Chamberlin, J. (2012). The work of experience based experts. In: J. Read & J. Dillon (Eds.), *Models of madness: Psychological, social and biological approaches to psychosis*, (pp. 305–319). Routledge: London.

Egner, J. (2018). 'The disability rights community was never mine': Neuroqueer disidentification. *Gender & Society*, *33*(1), 123–147.

Graby, S. (2015). Neurodiversity: Bridging the gap between the disabled people's movement and the mental health system survivors' movement? In: H. Spandler, J. Anderson & B. Sapey (Eds.), *Madness, distress and the politics of disablement* (pp. 231–244). Bristol: Policy Press.

Grey, F. (2017). Just borderline mad. *Asylum: The Magazine for Democratic Psychiatry*, *24*(1), 7–9.

Jones, N., & Kelly, T. (2015). Inconvenient complications: On the heterogeneities of madness and their relationship to disability. In: H. Spandler, J. Anderson & B. Sapey (Eds.), *Madness, distress and the politics of disablement* (pp. 43–55). Bristol: Policy Press.

McWade, B., Milton, D., & Beresford, P. (2015) Mad studies and neurodiversity: a dialogue. *Disability & Society*, *30*(2), 305–309.

Milne, E., Dickinson, A., & Smith, R. (2017). Adults with autism spectrum conditions experience increased levels of anomalous perception. *PLOS ONE*, *12*(5), p.e0177804.

Milton, D., and Timimi, S. (2016). https://blogs.exeter.ac.uk/exploringdiagnosis/debates/debate-1/ (Accessed 1 November 2019).

Plumb, A. (2015). UN Convention on the Rights of Persons with Disabilities: out of the frying pan into the fire? Mental health service users and survivors aligning with the disability movement. In: H. Spandler, J. Anderson, & B. Sapey (Eds.), *Madness, distress and the politics of disablement* (pp. 183–198). Bristol: Policy Press.

Recovery in the Bin (2019). Recovery in the Bin (blog). https://recoveryinthebin.org/. (Accessed 1 November 2019).

Romme, M., Escher, S., Dillon, J., Corstens, D., & Morris, M. (2009). *Living with voices: 50 stories of recovery*. Ross-on-Wye: PCCS Books.

Runswick-Cole, K. (2014). 'Us' and 'them'? The limits and possibilities of a politics of neurodiversity in neoliberal times. *Disability & Society*, *29*(7), 1117–1129.

Sapey, B. (2011). Which model of disability can include voice hearing experiences? In distress or disability? In J. Anderson, B. Sapey & H. Spandler (Eds.), *Proceedings of a symposium held at Lancaster University* (pp. 49–51). Centre for Disability Research, Lancaster University.

Spandler, H., & Barker M. J. (2016, 1 July). Mad and Queer Studies: interconnections and tensions. https://madstudies2014.wordpress.com/2016/07/01/mad-and-queer-studies-interconnections-and-tensions/ (Accessed 1 November 2019).

Spandler, H., & Poursanidou, D. (2019). Who is included in the Mad Studies Project? *The Journal of Ethics in Mental Health, 10.*

The Crisses. (2014). Longer Biography. https://kinhost.org/Crisses/LongerBiography (Accessed 1 November 2019).

The Redwoods (2018a). We are distinct people living in one body https://medium.com/@theredwoods/we-are-distinct-people-living-in-one-body-7ba6c13e6da8 (Accessed 1 November 2019).

The Redwoods (2018b). Many authors, one body. https://medium.com/@theredwoods/many-authors-one-body-6532311b5406 [Accessed 1 November 2019].

Timimi, S. (2018). The scientism of autism. *Mad in America.* www.madinamerica.com/2018/04/the-scientism-of-autism/ (Accessed 1 November 2019).

Walker, N. (2015). Neuroqueer: an introduction. https://neurocosmopolitanism.com/neuroqueer-an-introduction/ (Accessed 1 November 2019).

Woods, A. (2013). The voice-hearer. *Journal of Mental Health, 22*(3), 263–270.

Woods, A., Hart, A., & Spandler, H. (2019). The recovery narrative: politics and possibilities of a genre *Culture, Medicine, and Psychiatry,* 21 March.

Chapter 16

Neurodiversity studies
Proposing a new field of inquiry

Hanna Bertilsdotter Rosqvist, Anna Stenning,
and Nick Chown

This is not a book about the different forms of neurodivergence – although we hope that it will help to pave the way for autoethnographic writing that contradicts stereotyped or misleading understandings of neurodivergence. The editors share a commitment to creating 'theoretical' and methodological underpinnings that will help to make all our voices heard. It is hoped that this book contributes to the burgeoning cultural imperative in the West to redefine what it means to be human in the light of the experiences of neurodivergent people (as described in the Introduction).

Following ideas from black feminist theory, of 'centralizing marginality, marginalising the center' (Holling, 2018; hooks, 1984), as a way of seeing, or with other words, 'a way of approaching – what we do and how we do it' (Holling, 2018), we argue that centralising marginality, and marginalising the centre can be developed as a method for turning existing scholarship on autism and other forms of neurodivergence on their head. At this time, critical autism studies is at the centre of academic neurodiversity studies in the UK – this conclusion aims to decentre that, as neuroqueer theory has in the US. Where we focus on the level of experiences as present in phenomenology, others might seek affinities with other identity movements. The common goal is undermining the epistemic violence that says only a 'neurotypical' may determine the validity of our experiences and identities, and the actual violence that is often dependent on stigmatising particular identities.

We believe that through the focus on questioning hegemonic depictions of cognitive normality (curing neurodivergence/eugenics); shifting our focus onto neurodivergent wellbeing rather than pathology; and exploring cross-neurotype communication, from the position of neurodivergence, we contribute to the second orientation of neurodiversity theorisations, including the intricate task of challenging sometimes well-meaning 'doing NT business as usual' – both within research and in practice. It also includes *to see/look back* and *to talk back* to power by both questioning epistemic norms and practices, and developing new emancipatory methodologies. To be neurodivergent does not mean to automatically 'do' something or be informed by a neurodivergent perspective. We are all living in predominantly cognitive normative ideologies centring the experiences and perspectives of the neurotypical, and marginalising the experiences and perspectives of the neurodivergent. But we can start unlearning the cognitive normative gaze, as a way of cognitive decolonialisation. The consequences of this will continue to be registered in neurodiversity at work and in the new models of the mind posed by enactivism. While this book tends towards the 'theoretical' rather than the practical implications of neurodiversity, we hope others will continue to explore the ways that shallow or tokenistic understandings of neurodiversity risk marginalising us even further.

In discussions around what to include or exclude in neurodiversity, it is central for us to start to think in terms of both neurodivergence support perspectives and social identity perspectives. From a pragmatic neurodivergence support perspective, focusing on specific 'medical model' impairments as defined by individuals rather than imposed from outside, it is essential for people to continue receiving the support their lives depend on. If we lived in a 'universal design' society – in which the emphasis were on society to adapt to individuals – rather than vice versa, it may not be as necessary for people to disclose or even to diagnose their neurodivergence. However, we would still need to consider the likelihood of continued masking and oppression around other intersecting forms of identity, and the need to recognise 'co-occurring' conditions. Reasonable adjustments would be made automatically, and the stigma attached to some forms of neurodivergence would disappear. But we don't live in such a society, which makes the *centring* of neurodivergent needs based on neurodivergent rather than cognitive normate understandings of these needs absolutely essential.

From a social identity perspective, the matter of disclosure is complex and challenging for neurodivergent individuals given the stigma still attached to neurodivergence. In some cases, an individual will consider strategic disclosure to ensure they have support, legal protection from discrimination, and can avoid the stigma attached to public disclosure. While some people can disclose strategically, the stigma attached to some types of neurodivergence remains so great that disclosure may be too great a risk, particularly in some professions and locations. While autism is increasingly recognised as 'marketable' to employers, particularly in STEM subjects (Science, Technology, Engineering, and Mathematics), it is still understood as a deficit that renders certain forms of imaginative, communicative, and emotional labour impossible (as James McGrath (2017) has explored in *Naming Adult Autism*). It is even harder for ADHDers, Dyslexics and those with Tourettes to be included on the academic discussions of neurodiversity because of similar stigmatisation. For this reason, it is essential for neurodiversity studies to consider all neurodivergent differences and not to exclude types of neurodivergence that are seen as less culturally palatable. Rather, we want to argue that it is the responsibility of society to find a way to accommodate all variations, and for scholars to explore them rather than dismiss them.

The neurodiversity paradigm(s) can be seen as perspectives; either as lived experiences, ways of producing knowledge, ways of looking and talking back to power – of 'queering' the cognitive normative gaze – and an ethical stance. It is something different to what can be referred to as an 'allistic' perspective or 'NT business as usual'. This stems from the idea that knowledges of neurodiversity differ, depending on who one is talking about what, from what perspective, and in what context. Much of previous research on neurodivergent groups of people stems from a 'neurotypical-gaze'–position – regardless of the identity of the researcher. This means it is not just a matter of using certain theories in a particular data set, but doing research informed by a certain mind set; aiming at centralising cognitive marginality, and marginalising the cognitive centre. This approach is based on the assumption that 'normality' is a construct that harms everyone, especially, but not only, the neurodivergent. This also means that we need to recognise the possible tension between 'affirming lived experience' and 'queering'. We don't think that a tension is inevitable, rather that both can enable and inform each other. There is both a need to recognise neurodivergent lives and experiences of the world and to challenge cognitive norms.

Somewhere along the way, neurodiversity in an academic context has, up to a point, become coextensive with critical autism studies. We want to stress the importance of unpicking this, given that this orientation only addresses a limited range of neurodivergent experiences. One way of achieving this is to introduce the idea of biological citizenship, which was first used by Andriana Petryna (2002) in relation to rights and responsibilities of a state towards individuals with particular 'impairments'– and in the more critical sense, as used by Rose and Novas (2005). It may be used as a tool through which to question the ways particular bodies have been privileged through biopower. Neuro-equality feeds more into the original meaning and challenges the notion of rights based solely on conventional ideas of what is regarded as properly functioning cognitive capacities. This means not just finding out how to accommodate neurodivergent forms of life and practices but ensuring the full inclusion of neurodivergent people in collective decision-making.

This also brings out some new ontological and epistemological issues. Different interpretations of neurodiversity are referred to in this book – some more inclined towards medical formulations and the provision of support, and some toward social identities. What we share is an ideologically based assumption of human neurological diversity: following in the footsteps of queer theory, this means to counter the imaginary ideal of a cognitive normal subject. This includes challenging dominant notions of what it is to be human rather than to be disabled.

References

Holling, M. A. (2018). Centralizing marginality, marginalizing the center in the WSCA 2018 presidential address. *Western Journal of Communication*, 82(5), 529–536.
hooks, b. (1984). *Feminist theory: From margin to center*. Boston, MA: South End Press.

McGrath, James. (2017). *Naming adult autism: Culture, science, identity.* London: Rowman & Littlefield.

Petryna, Adriana. (2002) *Life Exposed: Biological citizens after Chernobyl.* Princeton, NJ: Princeton University Press.

Rose, N., & Novas, C. 2005. Biological citizenship. In A. Ong & S. Collier (Eds.), *Global assemblages: Technology, politics and ethics as anthropological problems* (pp. 439–463). Malden, MA: Blackwell.

Index

Page numbers in *italics* denote figures. Note entries are represented by 'n'.

Printed in the United States
by Baker & Taylor Publisher Services